AA003391

Computational Particle Technology

Topical Conference at the 2009 AIChE Annual Meeting

Nashville, Tennessee, USA
8-13 November 2009

ISBN: 978-1-61567-915-7

Printed from e-media with permission by:

Curran Associates, Inc.
57 Morehouse Lane
Red Hook, NY 12571
www.proceedings.com

Some format issues inherent in the e-media version may also appear in this print version.

Printed by Curran Associates, Inc. (2010)

AIChE
3 Park Avenue
New York, NY 10016-5991

Phone: (203) 702-7660
Fax: (203) 775-5177

www.aiche.org

TABLE OF CONTENTS

ANALYSIS OF DYNAMIC FLOW OF COHESIVE PARTICLES

Suresh Ahuja

ABSTRACT

Particulate flow of granular materials is usually such that cohesion is not negligible. Inter-particle forces, including van der Waals forces, capillary (liquid bridge) forces, sintering, electrostatic forces, and frictional forces contribute towards the properties of particles and aggregates in hoppers, risers, packed and fluidized beds and in pneumatic conveying. In shear flow of uniformly sized particles, inter-particle forces and hydrodynamic forces dominate. The properties of the system depend on two dimensionless numbers. The first one is related to the shear state imposed to the material. The second one characterizes the intensity of the cohesion. the macroscopic angle of friction at the peak state can be split into three distinct terms of collisional, frictional and dilational origins. In a square well cohesive model, different regimes are present depending on the magnitude of cohesive forces and solid fraction. Agglomeration from inelastic collisions (elastic-plastic) is simulated by decrease in number of particles but increase in size of the particles affecting frequency of inter-particle collisions. Shear stresses and normal stresses are compared against experiments in a shear cell (Freeman tester) using particles of different diameters, cohesive strength and dynamic friction.

INTRODUCTION

Cohesive inter-particle forces may affect powder flow, flow velocities and transition between a static granular medium and the avalanche process. Inter-particle forces include van der Waals, liquid-briding, electrostatic, and magnetic forces. The maximum stability angle is found to increase with the powder cohesiveness[1,2]. Numerical simulations, such as Particle Dynamics (PD) simulations, have been developed in order to go beyond the continuum analysis and take into account the discreteness of granular matter. In a PD Simulation, the bulk flow of the material is captured via simultaneous integration of the interaction forces between the individual particles Discrete models may involve particle to particle interactions as those between hard spheres or between soft spheres. The coupling of the DPM with a finite volume description of the gas-phase based on the Navier–Stokes equations has been done for the soft-sphere model and the hard sphere model respectively[3].

In a hard-sphere system the trajectories of the particles are determined by momentum-conserving binary collisions. The interactions between particles are assumed to be pair-wise additive and instantaneous. In the simulation, the collisions are processed one by one according to the order in which the events occur. Occurrence of multiple collisions at the same instant are not accounted in the model[4]. Additionally hard sphere approach fails in inelastic collisions particularly when high particle number densities or low coefficients of normal restitution e, the collisions will lead to a significant decrease in kinetic energy. Soft-sphere models use a fixed time step and consequently the particles are allowed to overlap slightly. The contact forces are subsequently calculated from the deformation history of the contact using a contact force scheme. The soft-sphere models allow for multiple particle overlap although the net contact force is obtained from the addition of all pair-wise interactions. The soft-sphere models are essentially time driven, where the time step should be carefully chosen in the calculation of the contact forces [5]. A DPM model based on the theory of contact mechanics was developed by Kafui, which enabled collision of the particles to be directly specified in terms of material properties such as friction, elasticity, elasto-plasticity and auto-adhesion [6]. In a square well model used for collisions between cohesive particles, several possibilities occur when the distance between the two particles approach twice the radius of the particle. Aggregation or separation after collision depends on relative normal velocity leading to capture-cohesive interaction (aggregation) or escaping-cohesive interaction. Compared to Hamaker's model where cohesive interactions lead to sticking of particles for a long time, square-well model particles bounce between physical contact and cohesive well. On incorporating cohesive forces into hard-sphere (molecular dynamics) simulations via a square-well potential, Weber etal. showed that cohesion attenuates the magnitude of the stress components at higher solids fractions (in the collisional regime) as compared to the non-cohesive case. At lower solids fractions (kinetic regime), however the presence of cohesive forces has little impact on the observed stress [7]. Cohesive forces were implemented in a discrete-particle fluidized bed simulation using both a square-well potential and a Hamaker model for van der Waals forces. For the parameters investigated, the results show that cohesive forces are significant in the pressure overshoot observed during the fluidization process. Furthermore, the mechanisms by which the specific cohesive mechanisms causing the pressure overshoot are different depending on the cohesion model being used. Although both models indicate that particle-particle cohesion dominates the overshoot, the square-well model predicts that cohesive interactions between particles and distributor plate also play a role, while the Hamaker model for van der Waals forces predicts that particle-sidewall friction may be enhanced by the presence of cohesion [8].

Using a model simulation of a simple flow geometry, the homogeneous plane shear, and the use of dimensional analysis, Pierre G. Rognon etal. found that constitutive law can be expressed by a simple friction law but that the cohesion strongly enhances the resistance to the shear. For small cohesion, due to the increase of the apparent friction between grains, the proportion of sliding contacts decreases which induces expansion of the material. For larger cohesion, the agglomeration of the grains results in the growth of heterogeneities (large voids separating dense granular areas), and in the increase of the contact force anisotropy, which strongly enhances the resistance to the shear. Then, for larger cohesion, the granular material breaks apart [9].

LARGE SCALE INTERACTIONS

Continuum approaches frequently describe granular materials using elasto-plastic models in which the material can support shear stresses up to a certain value before yielding and deforming irreversibly. In the simplest case, this approach amounts to Coulomb's law of friction. The most common yield criterion is the Mohr-Coulomb criterion, which proposes a linear relationship between the magnitudes of normal stress and shear stress in two-dimensions. Because the choice of axes dictates the relative magnitude of the normal and shear stresses, the range of possible axes maps out a circle in the space of normal and shear stress — the
"Mohr circle". The diameter of this circle is proportional to the magnitude of the stresses. In the normal-shear stress space, the Mohr-Coulomb yield criterion is the straight line $\tau = \sigma\mu + c$ where μ is the coefficient of friction and c is the cohesion. A rheological equation linking shear stress σ_{xy}, to normal stress σ_{yy} can be analyzed as

$$\sigma_{xy} = \mu\sigma_{yy} + \sigma_s + \mu\dot{\gamma}$$

where cohesive stress is given by σ_s viscosity by μ and shear rate $\dot{\gamma}$.

When μ = 0 this corresponds to a Bingham rheology. When σ_s =0, powder flow observes cohesive Coulomb rheology. If μ = 0 and σ_s =0, powder flow follows Newtonian rheology. In steady state flow the principal

stresses are proportional to each other. Using Mohr-Coulomb theory effective macroscopic friction coefficient of the powder material, which is the ratio of the shear stress to the normal stress at a shear plane can be determined [10].

Shear Testers

There are shear testers used to analyze powder flow namely, Jenike shear cell, Sevilla PowderTester and The FT4 Powder Rheometer. Jenike shear cell, has usually been used to test granular particles in pharmacetical industry. Distinct element simulation methods (DEM) and contact dynamics (CD) have been used to analyze influence of particle elasticity for non-cohesive granular matter in shear testers [11]. Jenike shear cell has been particularly applied to a study of flow patterns and stress distribution in silos. Limitation of a Jenike Shear Cell lies in showing non-uniform stress distribution and vertical force is limited to maximum of about 200psi. For cohesive particles in nano and micron size, clogging takes place. Sevilla PowderTester is a fluidized bed apparatus that enables one to test the bulk mechanical properties such as the yield stresses and compressibility of fine cohesive powders. Driving the powder into the bubbling regime in which the material loses memory of its previous history precedes every measurement. Then the gas flow is set to a given value to take the powder into a well-defined and reproducible initial state of low consolidation. The tester can be used to determine bulk mechanical properties such as the yield stresses and compressibility of fine cohesive powders. The testing involves several steps: Remove previous history by driving the powder into the bubbling regime, Gas flow is set to a given value to take the powder into a well-defined and reproducible initial state of low consolidation. Reverse flow is used to exert high compressive stress. After initializing, shear stress is applied to the sample by slowly tilting the bed. As the angle of tilt increases, this generates a shear stress in the powder layer. From the angle at which the sample fails in shear, one calculates the coordinates of one point on the yield locus. In order to generate more data for the yield locus we need to be able to vary the compressive stress on a sample. After initializing, shear stress is applied to the sample by slowly tilting the bed. As the angle of tilt increases, this generates a shear stress in the powder layer. For deep beds, the location of the shear plane and the angle of tilt at which failure occurs depend on the width of the bed [12]. The FT4 Powder Rheometer involves several steps A conditioning cycle comprises of the dynamic test blade slicing downward through the powder followed by an upward traverse that lifts the powder and drops it over the blade. This process helps to remove the effect of different sampling methodologies and powder storage times. The conditioned column of powder is compacted using a ventilated compaction piston (allowing entrained air to escape), with a force equal to that of the pre-shear normal stress. Critical consolidation is obtained by achieving incipient failure and the specimen is over-consolidated with respect to the normal stress applied during shearing. Reaching a critical consolidation level at steady state flow and then reducing the normal stress for shearing realize this. For each pre-shear normal stress, five measurements are taken at the five normal stresses defined by the standard. A measurement of shear stress is also taken at the preshear normal stress level i.e. at 3, 6, 9 and 15kPa. The five measurements taken make up the yield loci for each pre-shear normal stress [13]

A number of materials were put in F14 Freeman Tester and their shear stresses and normal stresses were computed after testing. An example of dependence of shear stresses on normal stresses for two size beads are given in Figure 1. Using the equation $\sigma_{xy} = \mu\sigma_{yy} + \sigma_s + \mu\dot{\gamma}$ at low shear rates $\sigma_{xy} = \mu\sigma_{yy} + \sigma_s$, μ,coefficient of friction and cohesive strength are calculated and are given in Table1.

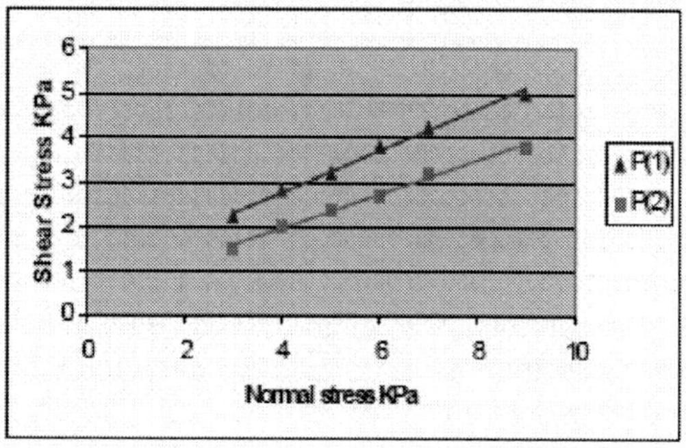

Figure 1:Two types particles P1 and P2 were subjected to powder flow in a shear cell of FT4.

Table 1: Dependence of particle size and coating on cohesive strength and coefficient of friction

Size of Particle/core surface	Polymer coating	Cohesive strength	Coefficient of friction
65 Micron steel	PMMA	0.83	0.47
35 Micron ferrite	Melamine/Kynar/PMMA	0.4	0.39
65 Micron steel	PMMA	0.82	0.47
35 Micron ferrite	Melamine/Kynar/PMMA	0.37	0.4

CONCLUSION:

Analysis of dynamic flow of cohesive particles is reviewed. The flows in the models are idealized and deviate from real practical flows. Experiments in particle flow also suffer and materials require certain modifications, fluidization etc. before cohesive flows can be tested. Strong dependence of particle size on cohesive strength consistent with model predictions is observed.

References

1 M. Valverde, A. Castellanos, and A. Ramos, (2000). Phys. Rev.E 62, p6851
2 C. Ursini, F. Restagno, G. Gayvallet, and E. Charlaix, (2002) Phys. Rev. E 66, 021304.
3 N. Olivi-Tran, N. Fraysse, P. Girard, M. Ramonda, an D. Chatain, (2002) Eur. Phys. J. E 25,p 217
4 M.J.V Goldschmidt R Beetstra J.A.M Kuipers (2002)Chemical Engineering Science 57 (11),pp 2059–2075
5 Y.Q Feng, B.H., Xu, S.J Zhang., A.B Yu, P Zulli (2004) A.I.Ch.E. Journal 50 (8), pp1713–1728.
6 K.D Kafui, C Thornton, M.J Adams (2002). Chemical Engineering Science, 57 (12), pp2395–2410.
7 MW. Weber, D K. Hoffman, C M. Hrenya (2004), Granular Matter,6(4) pp239-254
8 M W. Weber and C M. Hrenya(2007), Powder Technology,177(3)pp 170-184
9 P G. Rognon, J Roux, M Naa¨ım, and F Chevoir(2008) ournal of Fluid Mechanics 596 pp 21-47
10 T. Unger, L. Brandel, D. E. Wolf, J. Kert´esz,(2002) Phys. Rev.E, **65(6)**:pp 061305
11 J. Schwedes(2003) Granular Matter, **5**,pp1–43
12 A. Castellanos, J.M Valverde, (2004) KONA, 22, pp66-81
13 J.Cooke and T. Freeman, (2005) Annual Meeting of AAPS

Multi-Scale Modeling of Particle Flows in Granulators

Ben Freireich, School of Mechanical Engineering, Purdue University, West Lafayette, IN

Jianfeng Li, School of Chemical Engineering, Purdue University, West Lafayette, IN

James D. Litster, Chemical Engineering and Industrial Pharmacy, Purdue University, West Lafayette, IN

Carl R. Wassgren, Purdue University, West Lafayette, IN

Several researchers have attempted to couple discrete element method (DEM) and population balance (PB) models to predict granulation process performance. These studies have attempted to combine the strengths of DEM for modeling particle-scale phenomena such as collision frequency and collision velocity distributions, with the strengths of PB models for modeling macro-scale granulation processes over longer time scales. Here we present a different approach to combining DEM and PB models. Particle trajectories taken from the DEM model are used to generate an intermediate compartment model of particle flow through the spray zone of a paddle granulator, which can then in turn be easily used within a PB model framework. The compartment model presents a simplified, but still accurate, description of the particle residence and circulation times in the spray zone. The development of the DEM model and corresponding compartment model will be presented.

See more of this Session: Dynamics and Modeling of Particulate Systems I
See more of this Group/Topical: Topical 4: Computational Particle Technology

Constitutive Modeling of Slow Flows of Dense Granular Assemblies

Jin Sun, Chemical engineering, Princeton University, Princeton, NJ

Sankaran Sundaresan, Chemical engineering, Princeton University, Princeton, NJ

Slow flows of dense granular materials are encountered widely in industries and in nature. Reliable constitutive models for the rheological behaviors of dense granular flows that are linked to particle-level properties are not yet available; as a result, quantitative prediction of the flow characteristics in such systems through solution of continuum models remains elusive for the most part. In this presentation, we will describe a dissipative plasticity model [1, 2] constructed to meet such prediction needs for dense quasi-static flows.

To link the constitutive model to particle-level properties, we perform micromechanical analysis of dense granular assemblies under different types of deformations, including isotropic compression, simple shear, and triaxial compression. Towards this end, the discrete element method (DEM) is used to simulate the motion of individual particles in deforming assemblies; the stress and kinematics information at the continuum level are then obtained through statistical averaging. The simulations also afford detailed information about internal variables such as the fabric tensor and the coordination number of particles involved in force chains, which characterize the evolution of the microstructure of the assembly. These results are then analyzed to first identify the internal variables that are natural candidates to describe the microstructure and stress evolution and then seek quantitative relations that form the basis for continuum rheological models.

The dissipative plasticity model is thus composed of a rate-independent stress constitutive equation and the pressure and stress ratio relations, which are expressed as functions of the fabric tensor and coordination number associated with the force chains. The microstructure is evolved by evolution equations for these internal variables and the stress state varies accordingly. The evolution equations are developed by simplifying the general frame-indifferent functional forms of the time derivatives of the microstructure variables, as used for anisotropic fluids [3] and dense suspensions [4]. We will show that the model is properly defined so that the material constants depending on particle-level properties can be calibrated by the steady and unsteady simply shear data; the calibrated model is able to describe the rheological behaviors under those conditions with a good agreement. We will further demonstrate that the model is capable of predicting the rheological behaviors under other complex shearing conditions or different loading paths. Specifically, comparisons between model predictions and DEM data will be given for flows under cyclic shear with small strain amplitude, for Reynolds dilatancy and for flows under triaxial compression and extension.

[1] J. D. Goddard. Dissipative materials as models of thixotropy and plasticity. Journal of Non-

Newtonian Fluid Mechanics, 14:141-160, 1984.

[2] J. D. Goddard. A dissipative anisotropic fluid model for non-colloidal particle dispersions. Journal of Fluid Mechanics, 568:1-17, 2006.

[3] G. L. Hand. A theory of anisotropic fluids. Journal of Fluid Mechanics, 13:33-46, 1962.

[4] J. J. Stickel, R. J. Phillips, and R. L. Powell. A constitutive model for microstructure and total stress in particulate suspensions. Journal of Rheology, 50:379-413, 2006.

See more of this Session: Dynamics and Modeling of Particulate Systems I
See more of this Group/Topical: Topical 4: Computational Particle Technology

Validated DEM Simulations of Wet Granular Matter in a Bladed Mixer

Stefan Radl, Institute for Process and Particle Engineering, Graz University of Technology, Graz, Austria

Daniel Brandl, Institute for Process and Particle Engineering, Graz University of Technology, Graz, Austria

Brenda Remy, Dept. of Chemical and Biochemical Engineering, Rutgers University, Piscataway, NJ

Benjamin Glasser, Dept. of Chemical and Biochemical Engineering, Rutgers University, Piscataway, NJ

Johannes G. Khinast, Research Center Pharmaceutical Engineering GmbH, Graz, Austria

Flow and aggregation of powders is of central interest for applications like powder mixing, granulation or drying in the pharmaceutical industry. Despite intensive research has been carried out in this field, the description of wet powder flow is still demanding. For example, the complex flow in equipment with rotating internals, e.g., bladed mixers, is poorly understood from a theoretical point of view.

In our work we use specialized simulation techniques and a sophisticated measurement system to obtain a deep insight into wet granular flow. We use the discrete element method (DEM) to simulate the flow and mixing in a bladed mixer. Sophisticated particle interaction models were used to study both dry and wet bulk solid flow behavior (Radl et al. 2009). Furthermore, we have significantly improved our experimental setup (Lekhal et al. 2006) and are the first groups that is able to perform particle image velocimetry (PIV) of wet granular matter under well defined process conditions (pressure, temperature, stirrer speed, etc.).

Our goals were the investigation of the effects stemming from (a) the relative particle size, (b) the effect of moisture content, and (c) the effect of impeller speed on the mean and fluctuating velocity fields in the mixer. Our ultimate goal was the monitoring of a drying process by means of a video imaging system to yield improved process control. Such a system does not exist yet and would significantly contribute to process analytical technology (PAT) for particle processes.

Both numerical and experimental results reveal that the particle size has a significant effect on the instantaneous flow pattern in the mixer. This is due to a "roll-and-lock" effect that is caused by the gaps between big particles, i.e., the flow field has a higher fluctuation amplitude for large particles compared to small particles. Furthermore, we show that an identical granular Bond number is not sufficient to predict cluster formation in wet granular matter. Also, our experimental results show that there is a clear transition in the flow pattern when the moisture content is increased. This transition, however, also depends on the stirrer speed, i.e., the shear rate. This knowledge can be used to monitor, control and optimize drying processes in industrial applications.

References

Lekhal, A., Conway, S.L., Glasser, B.J., Khinast, J.G., 2006. Characterization of granular flow of wet solids in a bladed mixer. Aiche Journal 52, 2757-2766.

Radl, S., Kalvoda, E., Glasser, B.J., Khinast, J.G., 2009. Mixing Characteristics of Wet Granular Matter in a Bladed Mixer. Powder Technology (submitted).

See more of this Session: Dynamics and Modeling of Particulate Systems I
See more of this Group/Topical: Topical 4: Computational Particle Technology

Radial Mixing in Granular Materials in Rotating Cylinders

Suman K. Hajra, Computational Science Division, U. S. Department of Energy, National Energy Technology Laboratory,, Pittsburgh, PA

Mehrdad Massoudi, Computational Science Division, U. S. Department of Energy, National Energy Technology Laboratory,, Pittsburgh, PA

D. V. Khakhar, Department of Chemical Engineering, Indian Institute of Technology, Bombay, Mumbai - 400076, India

Particle self-diffusion has a significant effect on granular mixing. Rock-cuttings, sand, cement, tablets, grain, coal, biomass etc are an example of granular materials. These materials are transported from one place to another using belts, conveyors, drilling fluids or pneumatically. Flow induced mixing is important in handling of these materials. Rotating cylinder is commonly used as a prototype to study the granular mixing. We study the radial mixing using same size but different colors of glass beads in a quasi-2D cylinder rotated in the continuous flow regime. In this regime a shallow surface layer of particles flows steadily while the rest of the material rotates as a solid body. A qualitative study of the mixing is carried out using digital photography. Radial number fraction profiles of the tracer particles are obtained by bulk sampling. Since mixing occurs only in the flowing layer, hence mixing is considered in terms of ``passes" defined as the number of times the entire material in the bed flows through the layer. Results indicate that the mixing per ``pass" decreases with increasing rotational speed, and increases with increasing particles size. The mixed state captured by digital photography and the measured radial concentration profiles are well described by a convective diffusion model, using diffusivity as a fitting parameter. Mixing in transverse direction is found faster than azimuthal direction. One dimensional diffusion equation is used and a good agreement is obtained.

See more of this Session: Dynamics and Modeling of Particulate Systems I
See more of this Group/Topical: Topical 4: Computational Particle Technology

Cohesive Flow and Agglomeration of Granular Material in a Rotating Drum

Avik Sarkar,School of Mechanical Engineering, Purdue University, West Lafayette, IN

Carl R. Wassgren, Purdue University, West Lafayette, IN

The flow regimes of free flowing granular material in a rotating drum have been well characterized in the literature (see, for example, Mellmann (2001)). Numerical studies of cohesive granular flows in a rotating drum have been more focused towards mixing; systematic investigations of changes in flow behaviour are seldom reported. This work reports regime shifts seen with simultaneously varying inter-particulate cohesion and drum rotation speed (Froude number) from discrete element method (DEM) simulations. A new methodology for in situ identification of agglomerates is proposed; the key idea being that in order to qualify as an agglomerate, inter-particulate bonds must survive the dispersive forces acting to break them apart. Agglomerate size distributions have been determined using this new algorithm. Mixing trends are also quantified for qualitative comparison with the existing literature.

References

Mellmann, J., 2001. The transverse motion of solids in rotating cylinders - forms of motion and transition behavior. Powder Technology, 118, 251-270.

See more of this Session: Dynamics and Modeling of Particulate Systems I
See more of this Group/Topical: Topical 4: Computational Particle Technology

Numerical Solution to the Shearing Flow of Granular Materials

Mehrdad Massoudi, Computational Science Division, U. S. Department of Energy, National Energy Technology Laboratory,, Pittsburgh, PA

D. V. Khakhar, Department of Chemical Engineering, Indian Institute of Technology, Bombay, Mumbai - 400076, India

Suman K. Hajra, Computational Science Division, U. S. Department of Energy, National Energy Technology Laboratory,, Pittsburgh, PA

Studying flows of granular materials has become a major research initiative. One of the major challenges facing the designers of coal gasification plants is to assure reliable and efficient movement of solid particles into and out of high-pressure, high-temperature processing units such as fluidized beds. The more traditional approach of continuum mechanics still is used in many instances. Presently, there is no complete and full understanding of the constitutive relations that are needed for closure. We study the shearing flow of granular materials between two horizontal flat plates. The material coefficient such as viscosity and normal stress coefficient are not constant. The governing equations are non-dimensionalized and the resulting system of non-linear differential equations is solved numerically.

See more of this Session: Dynamics and Modeling of Particulate Systems I
See more of this Group/Topical: Topical 4: Computational Particle Technology

Modeling and Numerical Investigations of Stretching Liquid Bridges

Pirooz Darabi, Department of Mechanical Engineering, University of British Columbia, Vancouver, BC, Canada

Tingwen Li, Mechanical Engineering, University of British Columbia, Vancouver, BC, Canada

Konstantin Pougatch, Mechanical Engineering, University of British Columbia, Vancouver, BC, Canada

Martha Salcudean, Mechanical Engineering, University of British Columbia, Vancouver, BC, Canada

Dana Grecov, Mechanical Engineering, University of British Columbia, Vancouver, BC, Canada

Rupture characterization of a liquid bridge connecting wetted solid particles and the distribution of a liquid bridge between the particles plays a key role in many industrial particle processing systems. This work is aimed to provide a fundamental understanding of stretching liquid bridges. The work was applied to fluid cokers by investigating coke particle interaction with liquid bitumen; coating of particles and bridge formation.

A simplified mathematical model and numerical simulation of Navier-Stokes equations have been used to predict the shape evolution, rupture distance, and liquid distribution of pendular liquid bridges between two spherical solid particles. In the simplified model, the problem has been modeled using a parabolic assumption for the bridge shape. It is also assumed that surface tension forces are dominant over viscous, inertia, and gravitational forces. For numerical simulations, a commercial computational fluid dynamics software - FLUENT - has been employed. Numerical simulations have been based on Navier-Stokes equations; therefore, viscous, inertia and gravitational forces along with the surface tension forces have been taken into consideration. The use of numerical investigation has the following advantages to the simplified model: 1) Capability to solve more general rupture problems and 2) Capability to study the impact of operating conditions and liquid properties on the dynamics of the liquid bridge.

Results obtained with both the simplified model and numerical simulation for rupture distance have been validated with the experimental data available in the literature. It has been observed that both methods and experimental data were in reasonable agreement, provided the non-dimensional numbers were within a certain range. In order to study the effects of liquid properties (i.e. surface tension and viscosity) and operating conditions (i.e. separation velocity and gravity) numerous computations have been performed. As a general trend, the numerical simulation results show that as the importance of viscous forces relative to the surface tension forces increases, the rupture distance increases. On the other hand, increasing the importance of gravitational effects with respect to the surface tension effects reduces the rupture distance. Moreover, the numerical investigations show that the distribution of a liquid bridge between the two particles is strongly affected by the relative importance of the viscous, inertial, gravitational, and surface tension forces. It has been found that the simplified model provides results comparable to the predictions of the numerical simulation with Bond numbers less than 0.1 and capillary numbers less than 10-3. The simplified model has significant limitations, and is not applicable for non-dimensional numbers outside a certain range.

In conclusion, the simplified model and numerical simulation of Navier-Stokes equations are complementary when modeling the rupture distance and liquid distribution of liquid bridges: if the key dimensionless numbers are within a certain range, the simplified model can be used; otherwise, numerical simulation of Navier-Stokes equations has to be performed. The models presented in this work can be used to predict the rupture and post-rupture evolution of liquid bridges; ultimately, their results can be incorporated in numerical investigations of wet particulate processing systems.

See more of this Session: Dynamics and Modeling of Particulate Systems I
See more of this Group/Topical: Topical 4: Computational Particle Technology

Heat Transfer in Flowing Granular Materials

Mehrdad Massoudi, Computational Science Division, U. S. Department of Energy, National Energy Technology Laboratory,, Pittsburgh, PA

D. V. Khakhar, Department of Chemical Engineering, Indian Institute of Technology, Bombay, Mumbai - 400076, India

Suman K. Hajra, Computational Science Division, U. S. Department of Energy, National Energy Technology Laboratory,, Pittsburgh, PA

Heat transfer plays a major role in storage, handling, and processing of bulk solids such as grains, powders, pharmaceutical products, fossil fuels, etc. An important problem with wide applications in chemical industry is the flow of particulate matter on an inclined surface. In this work, we consider the free surface flow of granular materials down an inclined plane. The surface of the inclined is heated and the effects of radiation heat transfer at the free surface are studied. It is assumed that the material behaves like a continuum, similar to a compressible non-Newtonian fluid where the effects of density gradients are incorporated in the stress tensor. For a fully developed flow the equations simplify to a system of three non-linear ordinary differential equations. The equations are made dimensionless and a parametric study is performed where the effects of various dimensionless numbers representing the effects of heat conduction, viscous dissipation, etc, are presented.

See more of this Session: Dynamics and Modeling of Particulate Systems II
See more of this Group/Topical: Topical 4: Computational Particle Technology

Mathematical Modeling of Fluid Energy Milling Based On a Stochastic Approach

Shuli Teng, Chemical Engineering, New Jersey Institute of Technology, Newark, NJ

Peng Wang, Chemical Engineering, New Jersey Institute of Technology, Newark, NJ

Linjie Zhu, Polymer Processing Institute, Newark, NJ

Ming-Wan Young, Polymer Processing Institute, Newark, NJ

Costas G. Gogos, Otto H. York Department of Chemical, Biological and Pharmaceutical Engineering, New Jersey Institute of Technology, Newark, NJ

Abstract

In this article, the stochastic method is used to simulate the grinding process in a Fluid Energy Mill: the product particle size distribution is the result of repeating elementary breakage events, i.e. , where is the row vector of the feed particle size distribution, is the row vector of the product particle size distribution, is the number of elementary steps, and is a matrix of transition probabilities representing the elementary breakage event. The matrix of transition probabilities can be related to the breakage rate function and the breakage distribution function of the elementary breakage event. A specially designed apparatus, named "Single-event Fluid Mill" is employed to estimate those two breakage functions of the elementary breakage event with adding a breakage rate correction factor . The classification effect is considered by defining a cut size under which the particle will not break any more. Using this strategy, the product particle size distribution can be calculated. The good consistency between the simulation and the experimental results indicates that this model is valid to quantitatively estimate the grinding performance in the Fluid Energy Mill.

See more of this Session: Dynamics and Modeling of Particulate Systems II
See more of this Group/Topical: Topical 4: Computational Particle Technology

Mass Conservative Solution to the Dispersed Phase Size Distribution Using the Least-Squares Spectral Elements Method

Zhengjie Zhu,Chemcial Engineering, Norwegian University of Science and Technology, Trondheim, Norway

Hugo A. Jakobsen, Chemical Engineering, Norwegian University of Science and Technology, Trondheim, Norway

Carlos A. Dorao, Energy and Process Engineering, Norwegian University of Science and Technology, Trondheim, Norway

Loss of mass of dispersed phase in the standard least-squares formulation of the population balance equation is observed. This loss of mass is caused by inexact conservation property of some breakage models, which might result in incorrect physical interpretations.

Investigations on the reason for the poor mass conservation quality of some breakage models has not yet been reported. To the authors' knowledge, no remedy or modification in attempting to fix this critical problem has been made, either. This points out the need to find a possible way that improves those breakage models' performances by a guaranteed mass conservation.

In this work a constraint least-squares spectral element method is developed which enforces the mass conservation. This is accomplished by adding an extra restriction in the dispersed phase continuity equation through the Lagrange multipliers strategy. This modified version of least-squares formulation has been tested and compared with the standard one in to simulate the two-phase flow passing through a 2D domain, and the Martinez-Bazan breakage kernel is employed in the population balance equation.

The discretized system resulting from applying the method to a two-phase population balance equation problem is symmetric and pesudopositive definite. Results obtained by the modified least-squares spectral element method show that the mass is conserved everywhere with high accuracy.

See more of this Session: Dynamics and Modeling of Particulate Systems II
See more of this Group/Topical: Topical 4: Computational Particle Technology

DEM Simulations of Cylindrical Particles - Contact Detection Algorithms and Validation Studies

Madhusudhan Kodam, Mechanical Engineering, Purdue University, West Lafayette, IN

Rahul Bharadwaj, Pfizer Inc., Groton, CT

Jennifer S. Curtis, University of Florida, Gainesville, FL

Bruno C. Hancock, Pfizer Inc., Groton, CT

Carl R. Wassgren, Purdue University, West Lafayette, IN

In this study, algorithms are presented for determining if two cylinders, or a cylinder and an infinite plane, are in contact. In all, ten contact scenarios are examined involving the cylinder components (i.e. faces, band, and edges). In addition to checking for contact, relations for determining the contact location, overlap, and unit normal vector are also presented. DEM implementations of these algorithms are compared to an analytical solution for the post-collision translational and angular velocity for a single cylinder. Comparisons are also made to experimental measurements of the fractional residence time of a cylindrical object in a friability tester, and the solid fraction of the random packing of cylinders in a container. Comparisons to glued sphere representations of a cylinder are also made.

See more of this Session: Dynamics and Modeling of Particulate Systems II
See more of this Group/Topical: Topical 4: Computational Particle Technology

Modeling Fluidized Beds for Application to Biomass Systems

Emmanuela Gavi, Chemical and Biological Engineering, Iowa State University, Ames, IA

Joshua B. Drake, Mechanical Engineering, Iowa State University, Ames, IA

Jian Min, Chemical and Biological Engineering, Iowa State University, Ames, IA

Theodore J. Heindel, Department of Mechanical Engineering, Iowa State University, Ames, IA

Rodney O. Fox, Chemical and Biological Engineering, Iowa State University, Ames, IA

The use of gasification to produce biorenewable forms of energy has recently regained popularity [1-2]. This process is typically carried out in fluidized beds as they offer advantages such as low pressure drop, approximately uniform temperature distributions, high heat and mass transfer, and the ability to fluidize many particle types of various sizes. Biomass is injected into a heated bed of inert material, such as sand, and undergoes gasification, the result of which is a flammable hydrocarbon gas. This product can be used on site or can be further processed, for example through Fisher-Tropsch synthesis.

Improving gasification efficiency in fluidized beds requires a detailed understanding of the fluidization behavior. Computational Fluid Dynamics (CFD) modeling is an important tool for improving our understanding, however extensive validation of CFD models is required. In fact, when biomass is modeled, several required model inputs are not known or not easily measured experimentally [3].

The aim of this work is to validate CFD models applied to single component biomass systems against experimental data obtained by means of X-ray Computed Tomography (CT). An important result of the validation process is the needed use of an "effective density" for biomass in the CFD model. In fact, if the mid-value of the density range provided by the manufacturer is employed, the bulk density of the solid phase in the bed is overestimated, because the irregular particle shape and particle porosity are not included in the CFD model, which treats biomass particles as solid spheres.

In previous work [4] on the same fluidized bed facility, the validation of CFD models applied to glass beads systems showed that qualitative agreement between experiments and simulations was very good and quantitative bed pressure drop comparisons were also excellent. Measured local time-averaged gas fractions were also compared to simulated values with the overall agreement being satisfactory.

The present work employs ground walnut shell or ground corncob as model biomass bed material, and fluidization is performed at a gas velocity twice the minimum fluidization velocity ($U_g = 2\ U_{mf}$).

Simulations are run on the computational grid and with the drag model (Syamlal-O'Brien) employed in [4]. In order to tune the computational model for specific application to biomass, several combinations of particle sphericity (ϕ) and coefficient of restitution (COR) are tested. It is observed that the bed height is well predicted with COR = 0.9 for ϕ = 0.8--1. For smaller ϕ, the predicted bed height is higher, consistent with the results of Deza et al. [3]. The results also suggest that the value of COR has a negligible effect on the predictions for biomass systems.

References

1. Bridgewater, A.V., "Renewable fuels and chemicals by thermal processing of biomass", Chemical Engineering Journal, 91: 87-102 (2003).

2. Cui, H., and Grace, J.R., "Fluidization of biomass particles: a review of experimental multiphase flow aspects", Chemical Engineering Science, 62: 45-55 (2007).

3. Deza, M., Battaglia, F., and Heindel, T.J., "A Validation Study for the Hydrodynamics of Biomass in a Fluidized Bed" in Proceedings of FEDSM2008: 2008 ASME Fluids Engineering Division Summer Conference, Jacksonville, FL: ASME Press, Paper FEDSM2008-55158 (2008).

4. Min, J., Drake, J.B., Heindel, T.J., Fox, R.O., "Experimental validation of CFD simulations of a lab-scale fluidized bed reactor with and without side-gas injection", AIChE Journal, submitted (2008).

See more of this Session: Dynamics and Modeling of Particulate Systems II
See more of this Group/Topical: Topical 4: Computational Particle Technology

The Diametrical Compression Test as a Strength Prediction Method for Pharmaceutical Compacts

Tuhin Sinha, School of Mechanical Engineering, Purdue University, West Lafayette, IN

Bruno C. Hancock, Pfizer Inc., Groton, CT

Jennifer Sinclair Curtis, University of Florida, Gainesville, FL

Rahul Bharadwaj, Pfizer Inc., Groton, CT

Carl R. Wassgren, Purdue University, West Lafayette, IN

The diametrical compression test (a.k.a. Brazilian disk test, hardness test) is a common testing technique used in the pharmaceutical industry. The test produces an indirect tensile failure of the compact and, hence, provides a measure of the maximum force that can be sustained by a tablet before fracture. Due to the ease of performing this test, it is widely used as a control parameter during tablet compaction in order to prepare tablets with an acceptable mechanical strength.

The Finite Element Method (FEM) is used to investigate the anisotropy and plasticity generated during tablet compaction on the stresses produced within compacts during diametrical compression. Simulation results show that cracks initiate differently on both faces of the tablets resulting in a non-uniform stress profile along the thickness of the tablet. A constant stress value is assumed in the Hertzian/Hondros analytical solution which is commonly used to evaluate the maximum tensile force. High speed video experiments of the diametrical compression test show the non-uniform cracking behavior observed in the FE simulations. The current study highlights the influence of material heterogeneity on tablet strength.

See more of this Session: Dynamics and Modeling of Particulate Systems II
See more of this Group/Topical: Topical 4: Computational Particle Technology

The Hydrodynamic Behaviour of Gasifiers and Combustors

Joao F. Conde, Chemical Engineering, Heriot Watt Univesrity, Edinburgh, United Kingdom

Raffaella Ocone, Chemical Engineering, Heriot Watt University, Edinburgh, United Kingdom

The constant increase in carbon emissions from fossil fuel combustion and the recent legislation enforcing CO2 reduction will call for fundamental changes in energy demand, together with changes in fuel supply and use. New technology will be needed for the use of alternative renewable fuels (as, for instance biomass and wastes). Gasification offers a number of advantages, and more fundamental research in this area is in demand.

A fundamental model, originally developed for the hydrodynamics of suspension in bubbling beds, is extended to include effects relevant in circulating fluidised bed gasifiers and combustors. In particular, we investigate the extent of the solid/gas contact that can be achieved when the rate of heat transfer is high.

See more of this Session: Dynamics and Modeling of Particulate Systems II
See more of this Group/Topical: Topical 4: Computational Particle Technology

Brownian Dynamics Simulations of Nanoparticle Dispersions in Polymer Solutions

Ileana C. Carpen, Department of Chemical Engineering, Tennessee Tech University, Cookeville, TN

Deepika R. Gollamandala, Department of Chemical Engineering, Tennessee Tech University, Cookeville, TN

Addition of nanosized particles to a polymer solution/matrix can lead to new and improved properties over conventional composites using larger filler particles. The quality of a composite material depends upon the degree of dispersion of the filler particles. The nanoparticles have a tendency to aggregate which negates any advantages associated with the nanosize. The available experimental procedures to quantify the nanoparticle dispersions are expensive and laborious [1]. Hence, we propose to solve these problems by studying the nanoparticle-polymer systems using computer simulations to find the factors that affect the nanoparticle dispersions in polymer solutions. In this work, we investigated systems of nanoparticles and polymer using Brownian dynamic simulations. The shape of the nanoparticles was considered to be spherical and the polymer architecture was modeled as bead rod chains. The degree of dispersion and agglomeration in the system were characterized by mean square displacement. The effects of various individual factors, including nanoparticle polymer interactions, polymer chain length, and relative volume fractions, on the dispersion of nanoparticles in the polymer matrix were analyzed and presented.

Reference:

1. J. B. Putman, "Measuring Surface Roughness to Calculate Filler Dispersion in a Polymer Sample", US patent reference 5974167.

See more of this Session: Dynamics and Modeling of Particulate Systems II
See more of this Group/Topical: Topical 4: Computational Particle Technology

Fluid-Particle Drag in Sheared Particle Configurations

William Holloway, Chemical Engineering, Princeton University, Princeton, NJ

Jin Sun, Chemical Engineering, Princeton University, Princeton, NJ

Sankaran Sundaresan, Chemical Engineering, Princeton University, Princeton, NJ

Flows of gas-solid suspensions in large devices are usually analyzed through continuum models that take the form of balance equations for mass, momentum, and fluctuating motion of particle and fluid phases. In the momentum balance equation for the particle phase the dominant forces acting on the particles are usually the gravitational and the fluid-particle drag forces, with the remaining terms such as the divergence of the particle phase stress contributing to a smaller extent in most of the flow domain. As a result, an accurate constitutive model for the fluid-particle drag force is particularly important to predict the flow behavior quantitatively, and has indeed been the subject of a large number of studies in the literature; e.g., see Hill et al. [1], Li & Kuipers [2] and the references cited therein. Quite remarkably, all of the literature studies on this subject have examined (either theoretically or experimentally) the fluid-particle drag force in essentially homogeneous and (nearly) isotropic particle assemblies. However, suspension flows commonly involve spatial inhomogeneities in particle volume fraction and/or anisotropy in the local microstructure (which can arise even in homogeneous systems when the particle phase undergoes deformation). Ten Cate & Sundaresan [3] examined the effect of spatial gradients in particle volume fraction on the drag force and found it to be a small correction, but observed that changes in the radial distribution function for the particles did have measurably large effects on the drag force.

In the present study we have examined the effect of shear in the particle phase on the fluid particle drag force in the Stokes flow regime. When the particle phase undergoes shear, anisotropy develops in the particle microstructure, and fluid flowing through such an assembly experiences a drag force that differs appreciably from that for an isotropic particle assembly (at the same particle volume fraction). The permeability tensor now is distinctly anisotropic, whose extent increases with particle volume fraction. Indeed, this correction to the drag brought about the anisotropy can be as large as or even larger than the contributions to the momentum balance equations arising from the divergence of the particle phase stress. This study exposes the need for constitutive models that account for the changes in the fluid-particle drag force due to the changes in particle microstructure produced by local particle velocity gradients. Such a contribution to the fluid-particle drag may also produce a sizeable change to predicted stability limits of a uniformly fluidized state.

References:

[1] Hill, R., Koch, D., Ladd, A. (2001). The first effects of fluid inertia on flows in ordered and random arrays of spheres. J. Fluid Mech. 448: 243-278.

[2] Li, J., Kuipers, J.A.M. (2003). Gas-particle interactions in dense gas-fluidized beds. Chem. Eng. Sci. 58: 711-718.

[3] Ten Cate, A., Sundaresan, S. (2006). Analysis of the Flow in Inhomogeneous Particle Beds Using the Spatially Averaged Two-Fluid Equations. Int. J. Multiphase Flow, 32: 106-131.

See more of this Session: Computational and Numerical Approaches to Particle Flow I
See more of this Group/Topical: Topical 4: Computational Particle Technology

COUPLING OF HYDRODYNAMICS AND CATALYTIC REACTION IN FCC RISER REACTORS

Rajesh Patel, Dawei Wang, Chao Zhu[*]
Department of Mechanical & Industrial Engineering, New Jersey Institute of Technology, Newark, NJ 07102

Abstract

Fluid catalytic cracking (FCC) riser reactors convert heavy petroleum oils into light hydrocarbon products. The conversion rates, in principle, are dependent upon the coupling of hydrodynamics and reaction kinetics in the gas-solids transport process. The catalysts in a riser, on the other hand, undergo an accelerating transport process or a continuously diluting process, which is heavily influenced by the non-uniform cracking characteristics along the riser reactor. Most research on FCC riser reactor so far used hydrodynamic model of plug flow and reaction model based on overall catalyst-to-oil ratio (CTO), which not only neglects the non-uniform characteristics of hydrodynamics and reactions along the riser but also oversimplifies the intrinsic coupling of multiphase flow hydrodynamics and the cracking kinetics. This work is focused on the axial distributions and mechanistic coupling of local transport and reaction characteristics of catalysts and oils along the FCC risers. A simple four-lump reaction scheme is adopted to simulate the cracking reactions and all variable in both hydrodynamics and reaction models locally averaged over riser cross-sections. Modeling predictions are satisfactorily compared with the yield pattern and exit temperature of industrial scale plant data. The axial distributions of the hydrodynamic properties of the gas-oil and catalyst are plotted to study the effects of reaction kinetics on the axial distributions of the flow properties. Comparison against plug-flow-based modeling approach indicates that, the traditional methods tend to underestimate the vacuum gas oil (VGO) conversion percentage and gasoline yield in the lower part of the riser.

Keywords: Fluid catalytic cracking, Multiphase hydrodynamics modeling, Cracking kinetics, Riser reactor

Introduction

Fluid catalytic cracking (FCC) is the primary conversion process in the petroleum refinery industry. The hydrocarbon feed is supplied from bottom; and it vaporizes coming in contact with hot catalyst coming from regenerator. The vaporized gas oil feed cracks down to lighter molecules and it travels upward along with catalyst against gravity. The velocity of both gas and catalyst increases due to vaporization and cracking. The heat of the reaction for the cracking reaction is supplied by the hot catalyst. In the meantime, coke, a key byproduct of cracking reactions, gets deposited on the catalyst surface, which causes the catalyst to loose its activity on cracking.

Efforts have been made to improve the process productivity since the invention of the FCC process over a half century ago. However, from the modeling point of view, a mechanistic understanding of FCC process in a riser reactor is far from complete. There are multiple

---

[*] Author to whom correspondence should be addressed: Tel: 1-(973) 642-7624; E-mail: chao.zhu@njit.edu

challenges to overcome, including the complicity of catalytic cracking reactions which involves various simultaneous reactions of multi-component feed, the dynamic heterogeneity in vapor-catalyst transport in risers, and the inherent coupling of cracking reactions and hydrodynamics. Most of the previous FCC modeling have employed oil vapor in a few lumps of various products (e.g., four-lump, ten-lump, fourteen-lump, *etc.*). Weekman & Nace (1970) proposed a three-lump cracking model, Yen *et al.* (1987) proposed a four lump and Jacob *et al.* (1976) developed a more detailed ten-lump kinetic model. Even though the FCC process in a riser has been intensively investigated over the last few decades, most of the models are based on de-coupled isothermal plug flow hydrodynamics and the reaction model whose reaction rates are only related to the overall CTO of the entire riser.

Many researcher showed a growing interest on coupling the multiphase flow hydrodynamics into the FCC reaction process e.g., Han & Chung, 2001; Subramanya *et al.*, 2005; Gupta *et al.*, 2007]. However in their studies, the complicated flow structure and hydrodynamics of the gas-solids flow in the riser reactor has not been fully described. In above literature, only one way coupling has been presented in which only the influence from cracking reaction on flow hydrodynamics has been considered. The influence of flow hydrodynamics on cracking reaction, however, has not been coupled into the modeling.

In summary, a generic mechanistic model for FCC process has been presented, which not only predicts non-uniform characteristics of hydrodynamics and reactions along the riser but also the reasonable description of mechanistic coupling of multiphase flow hydrodynamics and the cracking kinetics. In order to avoid the mathematical complexities, a simplified four-lump kinetic scheme is adopted into our model.

Modeling Methodology

In this study, we propose a one-dimensional steady-state modeling approach. In both hydrodynamics and reaction model all variables are locally averaged over riser cross-sections. For simplicity, the entrance and end effects of the riser are excluded in this model. It is assumed that, at the riser inlet, feed oil instantly vaporizes by the immediate contact with the hot catalysts from the regenerator. For the convenience of calculating the total energy inside the riser reactor, it is further assumed that the vapor and the catalyst are always in local thermal equilibrium. The wall of the riser is assumed to be adiabatic. Reactions are assumed to be one-step and complete. A simple four-lump reaction scheme (Lee *et al.*, 1989) is adopted to simulate the cracking reactions, as shown in Figure 1.

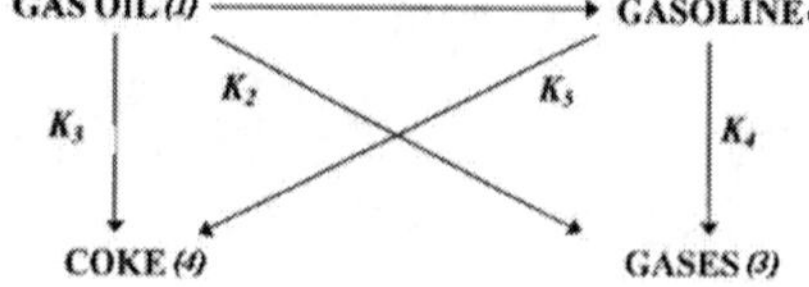

Fig. 1 Four-Lump model for gas oil cracking reactions (Lee *et al.*, 1989)

Plug Flow Model

We adopt the reaction scheme proposed by Ali *et al.* (1997) to simulate the cracking reactions. In our study, the equations are expressed with the weight fraction of hydrocarbons and

dimensional riser height instead of in a dimensionless form. The mass and energy balance for lumps in the plug flow model can thus be given as:
Gas oil:

$$\frac{dy_1}{dz}+\frac{\Phi_s A\alpha_g\rho_g}{F_g}\left[k_1+k_2+k_3\right]y_1^2\rho_g\alpha_g=0 \tag{1}$$

Gasoline:

$$\frac{dy_2}{dz}+\frac{\Phi_s A\alpha_g\rho_g}{F_g}\left[\left(k_5+k_4\right)y_2-k_1y_1^2\rho_g\alpha_g\right]=0 \tag{2}$$

Light Hydrocarbon Gases:

$$\frac{dy_3}{dz}+\frac{\Phi_s A\alpha_g\rho_g}{F_g}\left[k_4y_2+k_2y_1^2\rho_g\alpha_g\right]=0 \tag{3}$$

Coke:

$$\frac{dy_4}{dz}+\frac{\Phi_s A\alpha_g\rho_g}{F_g}\left[k_5y_2+k_3y_1^2\rho_g\alpha_g\right]=0 \tag{4}$$

Energy balance:

$$\frac{dT}{dz}+\frac{A\alpha_g\rho_g}{\left(G_sc_{ps}+G_gc_{pg}\right)}\left[y_1^2\rho_g\alpha_g(k_1\Delta H_1+k_2\Delta H_2+k_3\Delta H_3)+y_2(k_4\Delta H_4+k_5\Delta H_5)\right]=0 \tag{5}$$

Under the steady state condition, an energy balance between hot regenerated catalyst and gas oil feed at inlet is use to determine the temperature of the hydrocarbon mixture and catalyst at the riser inlet.

$$T_0=\frac{G_g.c_{pg}.T_B+G_s.c_{ps}.T_{s0}-G_g.\left(\Delta H_{fg}+m_g.c_{pl}\left(T_B-T_{l0}\right)\right)}{G_s.c_{ps}+G_g.c_{pg}} \tag{6}$$

Riser Flow Model

(A) Hydrodynamics Model

The overall mass balance of gas phase of all components is given by:

$$\frac{d\left(\alpha_g\rho_gU_g\right)}{dz}=-\left(r_3+r_5\right) \tag{7}$$

The average gas density can be calculated from the ideal gas law as:

$$\rho_g=\frac{P.\sum_{i=1}^{3}(C_iM_i)}{RT\sum_{i=1}^{3}C_i} \tag{8}$$

The mass balance of solid phase is given by:

$$\frac{d(\alpha_s\rho_sU_s)}{dz}=\left(r_3+r_5\right) \tag{9}$$

The momentum equation of gas phase can be expressed as:

$$-\frac{dP}{dz}=\alpha_g\rho_gg+\alpha_g\rho_gU_g\frac{dU_g}{dz}+F_D \tag{10}$$

The drag force per unit volume, F_D, is expressed by a modified Richard-Zaki equation (You *et al.*, 2009b) as:

$$F_D = \frac{18\mu}{d_s^2} \cdot \frac{\alpha_s}{(1-\alpha_s)^4} \cdot (U_g - U_s)\xi_1 \tag{10-a}$$

The momentum equation of solids phase can be expressed as:

$$F_D = \alpha_s \rho_s g + \alpha_s \rho_s g \frac{dU_s}{dz} + F_c \tag{11}$$

A semi-empirical model for the axial compact force (F_c) is proposed by (You *et al.*, 2009b):

$$F_c = (1 - k_2 k_3) F_D - (1 - k_3) \alpha_s \rho_s g \tag{11-a}$$

The energy balance between the hot catalyst and gas oil feed during the reaction is given by;

$$(\alpha_s \rho_s u_s c_{ps} + \alpha_g \rho_g u_g c_{pg}) \frac{dT}{dz} + (r_3 + r_5)(c_{ps} - c_{pg}) T = -\sum_{i=1}^{5} r_i . \Delta H_i \tag{12}$$

(B) Reaction Model

Based on the reaction scheme shown in Fig. 1, the reaction equation for each component of reactant is given by:

Gas-oil:

$$U_g \frac{dC_1}{dz} + C_1 \frac{dU_g}{dz} = -\Phi_s (k_1 + k_2 + k_3) C_1^2 \tag{13}$$

Gasoline:

$$U_g \frac{dC_2}{dz} + C_2 \frac{dU_g}{dz} = \Phi_s \left(\frac{M_1}{M_2} k_1 C_1^2 - (k_4 + k_5) C_2 \right) \tag{14}$$

Light Gases:

$$U_g \frac{dC_3}{dz} + C_3 \frac{dU_g}{dz} = \Phi_s \left(\frac{M_1}{M_3} k_2 C_1^2 + \frac{M_2}{M_3} k_4 C_2 \right) \tag{15}$$

Coke:

$$U_g \frac{dC_4}{dz} + C_4 \frac{dU_g}{dz} = \Phi_s \left(\frac{M_1}{M_4} k_3 C_1^2 + \frac{M_2}{M_4} k_5 C_2 \right) \tag{16}$$

The rate equation of the ith conversion can be described using the Arrhenius type expression:

$$k_i = k_{i0R} \lambda \exp\left(-\frac{E_{ai}}{RT_i} \right) \tag{16-a}$$

where k_{i0R} is pre-exponential factor for riser flow, which is derived from the plug flow model. The pre-exponential factor for riser flow is molar-based, which is expressed as:

$$K_{i0R} = \left(\frac{M_i}{\alpha_g \rho_g} \right) K_{i0} \tag{16-b}$$

λ represents the local catalyst to oil ratio, which was related to the overall mass flow ratio of catalyst to oil (CTO) or the feeding mass flow rate ratio of catalyst over hydrocarbon gases by:

$$\lambda = \left(\frac{\alpha_s}{\alpha_{s,avg}} \right)^n \cdot \left(\frac{C}{O} \right)_{overall} \tag{16-c}$$

The power index 'n' in Eq. (17-c) is selected as 0.25 in this study, obtained from data match against industrial plant measurements. Replacing the overall catalyst to oil ratio with the localized ratio, the cracking reaction rate can be directly related to local mass ratio of oil and catalyst instead of phase velocities only.

The parameter Φ_s represents the catalyst decay of activity due to coke deposition. In the present work, we used catalyst deactivation function proposed by Pitault *et al.* (1994)

$$\Phi_s = \frac{X+1}{X+\exp(Y.C_c)} \tag{16-d}$$

In summary, for a complete modeling description of FCC hydrodynamics coupled with reactions, we have obtained a total of 10 independent Eq. (7) to (16) for total 10 independent variables (α_s, U_s, U_g, C_1, C_2, C_3, C_4, ρ_g, P, T). Hence, the problem is closed.

Results and Discussion

The model predictions of the product yields and axial profiles of transport parameters are discussed in this section. The model predictions are validated against two sets of FCC plant data (Ali *et al.*, 1997; Derouin *et al.*, 1997). The operation parameters and physical properties of catalyst are lump components are listed in Table 1, while reaction constants and heat of reactions are listed in Table 2.

Table 1 Operation parameters and FCC properties of riser reactors

Operation parameters & FCC properties	**Case 1 (Ali & Rohani, 1997)**	**Case 2 (Derouin *et al.*, 1997)**
Catalyst feed rate (kg/s)	144	470
VGO feed rate (kg/s)/CTO	20/7.2	85/5.5
Inlet temperature of VGO feed (K)	496	650
Inlet temperature of catalyst (K)	960	960
Riser diameter (m)	0.8	1
Riser height (m)	33	35
Catalyst diameter (μm)	70	75
Inlet riser pressure (atm)	2.9	3.15

Table 2 Pre-exponential factor, Activation energy, and Reaction heat (Han & Chen, 2001)

Cracking Reaction	**ΔH_i (kJ/kg)**	**K_{i0} (s^{-1})**	**E_a (kJ/kmol)**
VGO → Gasoline	195	1457.5	57359
VGO → Light Gases	670	127.59	52754
VGO → Coke	745	1.98	31830
Gasoline → Light Gases	530	256.81	65733
Gasoline → Coke	690	0.022	66570

***3.3 Comparisons of Riser Flow Model conversions with Plug Flow Model* conversion and Plant data**

The comparison of model predictions of yields and conversion with plug flow model and FCC plant data of Ali *et al.* (1997) & Derouin *et al.* (1997) are shown in figure 3.

Figure 2(a) shows that both models predictions of various yields reasonably agree with the plant data at the exit of riser. However, in the dense transport regime, our model predictions shows a better agreement than those of the plug flow reaction model as shown in

Fig. 2(b), which is attributed to high local catalyst to oil ratio in this region. In the dilute phase transport regime, our riser model shows a more asymptotic trend of the gasoline yield, which agrees with the trend of reported plant data.

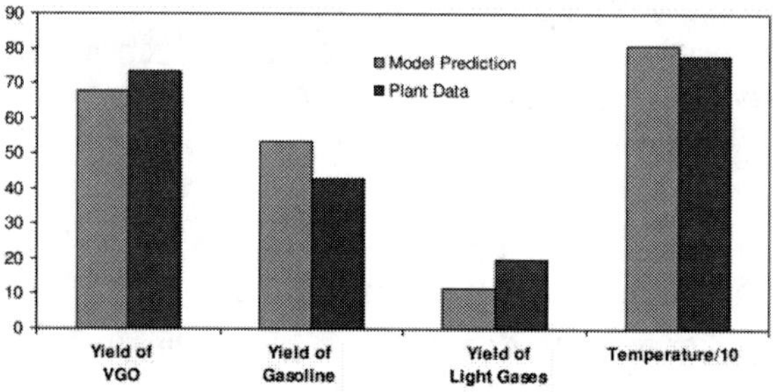

(a) FCC plant data of Ali *et al.*, 1997.

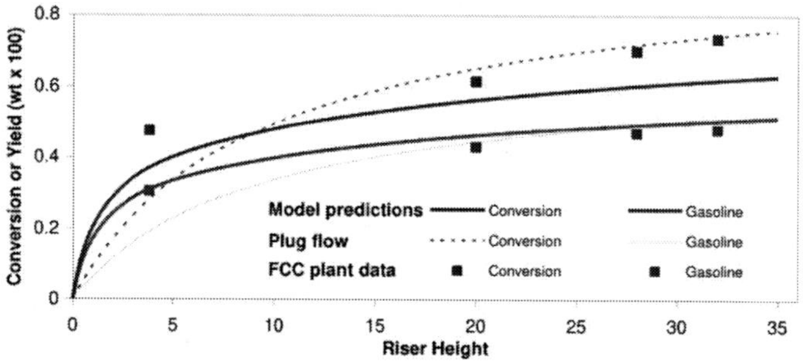

(b) FCC plant data of Derouin *et al.*, 1997.

Fig. 2 Comparison with plug flow model predictions.

3.4 Axial Profiles of Hydrodynamics and Reactions Parameters

The axial distribution of the hydrodynamic parameters like phase velocities, temperature and pressure are shown in figures (3), (4), and (5) using operation conditions of Case 2 in Table 1.

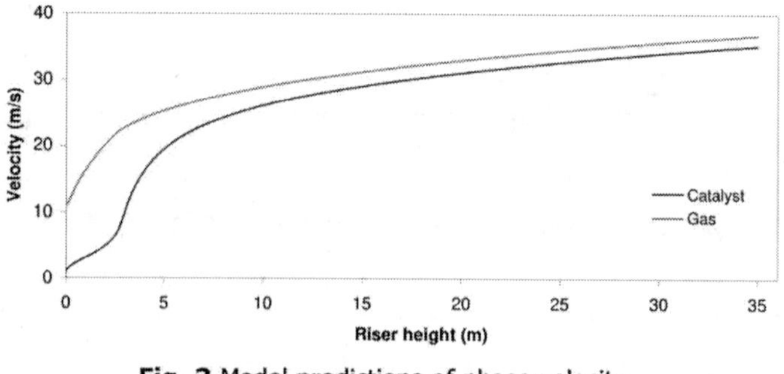

Fig. 3 Model predictions of phase velocity.

Figure 4 shows the model predictions of gas and solid phase velocities along the riser reactor. The gas phase velocity is increasing along the reactor height, which is due to the reaction-induced vapor moles expansion. However in the lower part of the riser, the increase in gas phase velocity is much larger than the rest of the riser; this is attributed to the intensive cracking reactions in bottom of riser. The velocity of solids phase follows the same trend of gas phase velocity, as the solids are carried by gas phase.

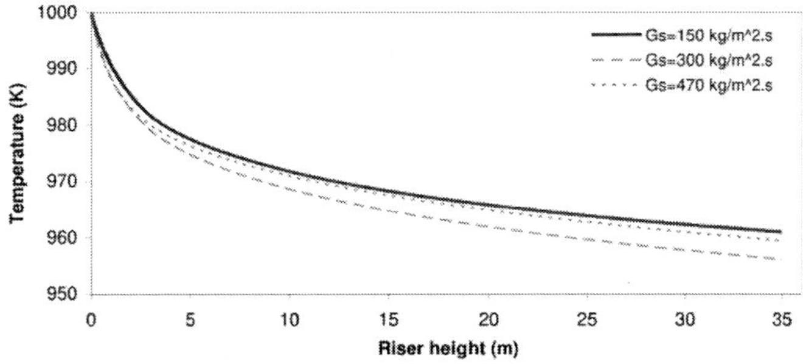

Fig. 4 Model prediction of riser temperature.

Figure 5 shows the axial profile of temperature with the assumption of local equilibrium between gas feed and catalyst. Due to the endothermic cracking reaction, the temperature drops monotonically along the riser. Since the reaction rates are strongly dependant upon temperature, the decrease in temperature slows down the cracking reactions and hence the gradient of temperature along the reactor height.

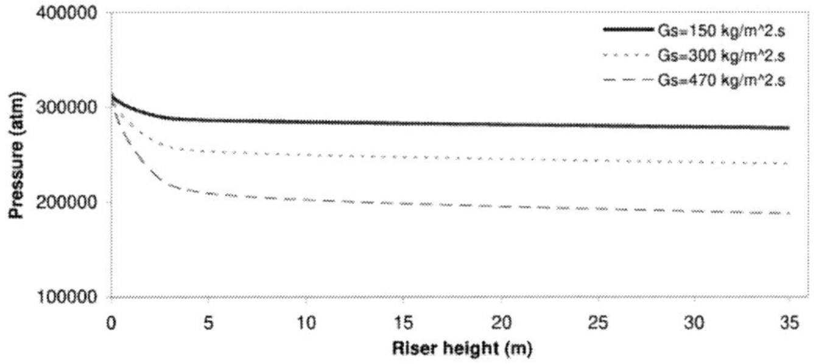

Fig. 5 Model prediction of riser pressure

Figure 5 shows the model prediction of axial distribution of pressure along the riser reactor, with a reasonable trend. The pressure decreases very fast in the dense phase region, partly due to the intensive inter-particle collisions induced energy dissipation, temperature drop due to endothermic cracking reaction, and kinetic energy for catalysts acceleration and dense phase transport. It is noticed that the back mixing of the solids in the riser flow is very significant, which results in crease in mass flow rates of solids in core region. The effect of solids back mixing is not included in this study, hence the pressure drop profile at riser exit differs from reality.

Conclusion

A mechanistic model is proposed in this paper, which describes the mechanism to consider the inter-coupling between gas-solid flow hydrodynamic and cracking reaction kinetics. The predictions from the proposed riser model have been reasonably validated against the two different sets of published FCC plant data. The model predictions are also compared with traditional reaction model whose hydrodynamics is based on plug flow. It shows that flow hydrodynamics has significant influence on reaction cracking kinetics, which has not evoked enough attention in the past. This modeling approach has successfully combined FCC reaction kinetics with local flow hydrodynamics, which will help to understand the influence of flow hydrodynamic on reaction kinetics and vice versa. The proposed modeling mechanisms may provide valuable insights of the FCC reaction, and hence benefit the optimization in reactor design and operation.

Nomenclatures

A	riser cross section area (m^2)	C	molar concentration of lump (mol/m^3)
D	riser diameter (m)	E_{ai}	activation energy of ith conversion (kJ/mol)
F	mass flow rate (kg/s)	G	mass flux rate (kg/m^2 s)
k_i	rate constant of ith conversion (s^{-1})	M	molecular weight (kg/mol)
P	pressure (atm)	R	universal gas constant (J/mol K)
T	temperature (K)	U	velocity (m/s)
y_i	weight fraction of ith lump	z	riser Height (m)
ΔH_i	heat of reaction from i^{th} component (J/kg)	ρ	density (kg/m^3)
α	volume fraction	μ	dynamic viscosity of fluid ($N.s/m^2$)
r_{ij}	mass rate change from i^{th} component to j^{th} component (Kg/m^3 s)		

Subscripts

s	solid	g	gas
0	inlet condition	p	particle

Reference

1. Weekman, V. W., Nace, D. M., "Kinetics of catalytic cracking selectivity in fixed, moving, and fluid bed reactors", *AIChE J.* 16 (3): 3397-3404, 1970.
2. Yen, L., Wrench, R., and Ong, A., "Reaction Kinetic Correlation for Predicting Coke Yield in Fluid Catalytic Cracking", Presented at the Katalistisks' 8th Annual Fluid Catalytic Cracking Symposium, Budapest, Hungary, June 1-4, 1987.
3. Jacob, S. M., Gross, B., Voltz, S. E., & Weekman, V. W., "A lumping and reaction scheme for catalytic cracking", *AIChE J.* 22: 701-713, 1976.
4. You, J., Patel R., Wang, D. W., Zhu C., "Role of Inter-particle Collision on Solids Acceleration in Riser". *J. of Particuology*, In press, 2009b.
5. Ali, H., Rohani, S., "Dynamic Modeling and Simulation of a Riser-Type Fluid Catalytic Cracking Unit", Chem. Eng. Technol., 20: 118-130, 1997.
6. Han, I. S., Chung, C. B., "Dynamic modeling and simulation of fluidized catalytic cracking process, Part-II: Property estimation and simulation", *Chem. Eng. Sci.* 56: 1973-1990, 2001.
7. Lee, L., Chen, Y., Huang, T., Pan, W., "Four lump kinetic model for fluid catalytic cracking process", *Can. J. of Chem. Eng.* 67: 615–619, 1989.

Non-Spherical Particle Modelling in Fluid Flow Using DEM-CFD Coupling

Wai Sam Wong, DEM Solutions Ltd., Edinburgh, United Kingdom

Richard D. LaRoche,DEM Solutions USA Inc., Lebanon, NH

In numerical models, particles are usually simplified and assumed to be spheres. However, particles are typically irregular shaped in many real-life industrial applications, such as sedimentation, mineral processing, agricultural and more. Most of these applications involve modelling of gas-solid flow. Thus, the dynamics of non-spherical particles in fluid is of great interest. Early investigations have been focussing on drag over spheres in fluids. For non-spherical particles, the drag and lift forces will be a function of its orientation against the fluid. Torque on the particles induced by the hydrodynamics forces will need to be included as well.

DEM-CFD coupling model has been an emerging approach in particle-fluid interactions modelling. It has been successfully applied in many gas-solid flow applications, such as pneumatic conveying and fluidised beds. In DEM, a method to utilise spheres to approximate the shapes of non-spherical particles and to model their dynamics is well-established. This method is employed in the leading commercial DEM software, EDEM. However, in gas-solid flow, DEM-CFD calculations of hydrodynamic forces and toques on non-spherical particles are often based on spherical assumptions. There are drag and torque models devised for non-spherical particles tracking in fluid, but these models mainly work well in uniform flow.

In this paper, the hydrodynamic models for non-spherical particles in uniform flow have been implemented. We also propose a simple model that would extend the original model to take into account the velocity gradient around a non-spherical particle in non-uniform flow, e.g. jet acting on part of a cylinder. The model has been validated against an experiment, where a cylinder is dropped into a rectangular fall tube. The model is then applied to an industrial application setup to demonstrate the potential of this model in flow with uniform and non-uniform fields.

See more of this Session: Computational and Numerical Approaches to Particle Flow I
See more of this Group/Topical: Topical 4: Computational Particle Technology

Simulation of Powder Flow in Die Filling Applications

Peter Spicka, Fluent Inc., Lebanon, NH

I.C Sinka, University of Leicester, Leicester, United Kingdom

Weight uniformity of tablets in die filling applications is proportional to the consistency of the powder stress distribution and the flow of the powder in the feed frame of a die fill. Therefore, understanding the stress distribution profiles within a feed frame appears to be the key factor for optimal design of the die fill in terms of feeder paddle geometry and compression speed. Recently, discrete element methods have been quite frequently used for powder flow application. Although they are rather accurate in predicting particle-particle interactions, the discrete element methods are severely limited by maximum number of tracked particles and thus are limited to a relatively small domain size. On the contrary, the Eulerian-based models are independent of the number of particles but the implementation of frictional forces is indirect and not well understood. Therefore, Lagrangian-based Particle-In-Cell techniques, which can be used to track a relatively small number of particles while maintaining the ability to model the contact forces directly, appear the most suitable choice for powder flow analysis. In the current study, we compare numerical predictions of Eulerian granular and Particle-In-Cell models that have been developed in commercial CFD software Fluent 12. Numerical results obtained by the models are validated against experimental data at different flow conditions.

See more of this Session: Computational and Numerical Approaches to Particle Flow I
See more of this Group/Topical: Topical 4: Computational Particle Technology

Solution of the Boltzmann Equation by the FCMOM

Matteo Strumendo, Department of Chemical and Biological Engineering, Illinois Institute of Technology, Chicago, IL

Hamid Arastoopour, Chemical and Biological Engineering, Illinois Institute of Technology, Chicago, IL

The fluid dynamics of the particulate phase in gas-solid flows (except in the regions where frictional forces are dominating) is described by the Enskog-Boltzmann equation for inelastic particles; in dilute systems, the Enskog-Boltzmann equation reduces to the Boltzmann equation. Several methods are available to solve the Boltzmann equation, i.e. generalizations of the Chapman-Enskog expansion, method of moments, Monte Carlo simulations, discretization of the Boltzmann equation in the velocity space, Lagrangian methods; in this work, the focus is on the solution of the Boltzmann equation by the method of moments.

The first application of the method of moments to the Boltzmann equation was developed by Grad [1], who solved the Boltzmann equation for simple (elastic) gases using a third order approximation of the particle velocity distribution. Strumendo and Canu [2] generalized the Grad approach and were able to compute the anisotropy of the granular flows, however their technique is not successful when extended to higher order moments.

The Boltzmann equation can be interpreted as a multi-variate population balance equation, in which the internal variables are the components of the particle velocities; more specifically, the Boltzmann for bi-dimensional systems (hard disks) is a bi-variate population balance equation, while the Boltzmann for tri-dimensional systems (hard spheres) is a tri-variate population balance equation. In this perspective, concepts and techniques developed in the solution of population balance equations can be used for the solution of the Boltzmann equation, and vice versa.

Recently, several variations of the method of moments (QMOM, DQMOM, FCMOM) were developed and were capable to provide numerically efficient and accurate solutions of population balance equations. It is interesting to investigate whether and to which extent such techniques can be applied for solving the Boltzmann equation.

Fox [3] explored the possibility of applying quadrature-based closures in the solution of the kinetic equation for dilute gas-particle flows, thus enlarging the spectrum of applications of some of the core ideas of the QMOM/DQMOM techniques.

In this work, the FCMOM technique is extended for solving the Boltzmann equation. The FCMOM

(Finite size domain Complete set of trial functions Method Of Moments) was developed by Strumendo and Arastoopour ([4], [5]) to solve mono-variate and bi-variate population balance equations. The fundamental idea is to construct a method of moments on a finite domain of the internal variables (typically the particle size in mono-variate population balance equations, or the particle velocities in the Boltzmann equation). An advantage of the FCMOM is that it provides both the moments and the reconstructed distribution of the particle internal variables. Further, the domain of the internal variables is always well defined (a property which is relevant in multi-variate applications).

The proposed method is applied to the Boltzmann equation for bi-dimensional systems (hard disks) and is illustrated through homogeneous and in-homogeneous applications. In the homogeneous case, the relaxation of elastic particles to the Maxwellian state, and the asymptotic behaviour of inelastic particles (Homogeneous Cooling State solution) are considered. In the in-homogeneous case, the impulsive start-up problem is considered.

[1] Grad, H., "On the Kinetic Theory of Rarified Gases", Communications on Pure and Applied Mathematics, 2, 331-407, (1949).

[2] Strumendo, M., Canu P., "Method of Moments for the Dilute Granular flow of Inelastic Spheres", Physical Review E 66, 041304/1-041304/20, (2002).

[3] Fox, R.O., "A quadrature-based third-order moment method for dilute gas-particle flows",

Journal of Computational Physics, Volume 227, Issue 12, 6313-6350, (2008).

[4] Strumendo, M., Arastoopour, H., "Solution of PBE by MOM in Finite Size Domains", Chemical Engineering Science 63, 2624-2640, (2008).

[5] Strumendo, M., Arastoopour, H., "Solution of Bivariate Population Balance Equations Using the FCMOM", Industrial and Engineering Chemistry Research 48(1), 262-273, (2009).

See more of this Session: Computational and Numerical Approaches to Particle Flow I
See more of this Group/Topical: Topical 4: Computational Particle Technology

2009 AIChE Annual Meeting
Nashville, TN
November 8-13, 2009

CPFD Eulerian-Lagrangian Method for Three Dimensional Thermal Reacting Flow An Example of a Syngas Gasifier

Dale Snider, CPFD Software, 10899 Montgomery NE, LLC, Albuquerque, NM 87111
Sam Clark, CPFD Software, 10899 Montgomery NE, LLC, Albuquerque, NM 87111

Introduction

This study provides a description of CPFD (Snider 2009) modeling for fluid-solid flow with solids and gas chemistry. The example chemistry is pyrolysis of solids and homogenous gas chemistry in a high temperature thermal environment. The example problem, illustrating the computational method, is a large three dimensional gasifier.

For the most part, mathematical models of separated particulate multiphase flow have used either a continuum approach for both phases (Gidaspow, 1986, Batchelor, 1988, Jackson, 2000) or a continuum approach for the fluid and a Lagrangian computational-particle model for particle phase (Cundall and Strack, 1979, Amsden, et al,, 1989, Snider 2001, Godlieb, et al, 2007). The two-fluid continuum approach averages both fluids and solids by a statistical procedure (typically ensemble-averaging) and treats the solids phase as a pseudo continuum. The averaging procedure leads to many unclosed terms which must be modeled (Jackson, 2000), and much progress has been made in the modeling of such terms as the particle-particle-stresses and particle viscous-stresses in dense particle flows using spatial gradients of average particle properties (Batchelor, 1988, Gidaspow, 1994). The two-fluid continuum approach has trouble modeling flows with a distribution of particle types and sizes because separate continuity and momentum equations must be solved for each size and type (Gidaspow, 1994, Risk, 1993). Size variation and different solid materials affect the bed dynamics with possible solids segregation, and the solid size distribution can produce or suppress gas flow patterns such as gas streaming (Karri, et al. 2004). In thermal, reacting flows, the solid not only changes size from pyrolysis, but the rate of reactions and temperatures can depend on solids surface area, type of solid material and discrete solids temperature.

Modeling the particle phase as discrete particles or elements has a number of variations. Direct numerical solution and Lattice Boltzmann calculations are limited to a small number of particles. Both methods are useful in studying fundamental solid-fluid flow behavior and can be used in developing closure models for fluid-particle interactions needed in continuum models. The discrete particle or discrete element method (DPM or DEM) (Cundal and Strack, 1979) models a finite number of discrete semi-rigid particles interacting through contact forces and transferring momentum to and from the fluid by a drag closure model. The multitude of particle-to-particle contacts and the model for the contact force are the most significant parts of the direct element model. To efficiently calculate particle interactions, relatively simple models, often based on a simple coefficient of restitution, are employed. DPM allows solution for flows with a wide range of particle types, sizes, shapes and velocities. Because, of the high collision frequency for volume fractions above 5% (O'Rourke, 1981) and the computational complexity of calculating dense particle-particle interactions, DPM calculations have been limited to on the order of $2x10^5$ particles and are often restricted to two-dimensional solutions without a fluid phase (Godlieb, et al., 2007).

This paper describes the CPFD approach to modeling thermal reacting fluid-solid flows. The CPFD numerical methodology incorporates the multiphase-particle-in-cell (MP-PIC) method (Andrews and O'Rourke 1996, Snider 2009, Snider et al 1998) for calculation particle physics. The MP-PIC method is a discrete particle method (DPM) where the fluid phase is treated as a continuum

(Eulerian) and the solids are modeled by Lagrangian computational particles. O'Rourke first proposed the MP-PIC method and demonstrated the method with one-dimensional simulations compared to analytical solutions and experimental data (Andrews and O'Rourke, 1996). The MP-PIC method is an extension of the stochastic particle method of the KIVA code (Amsden, et al, 1989) in which an isotropic particle stress gradient is added to the equation of motion of the particles. This addition enabled the method to calculate flows with any particle volume fraction from the dilute to the close pack limits. The particle stress gradient, which is difficult to calculate for each particle in dense flow, is calculated as a gradient on the grid, fully coupled to the other particle and gas acceleration terms, and is then interpolated to the discrete particles.

This paper focuses on the thermal and chemical modeling of chemistry with pyrolysis of the solid. Previously, Snider and Banerjee (2009) presented CPFD numerical methodology applied to isothermal, catalytic chemistry. The example problem illustrating the computational method is a large three dimensional gasifier. The bed has complex solids and gas flow patterns which produce gas-solid chemistry and gas-gas chemistry patterns. Coal gasification is expected to provide a versatile and clean method of converting coal into electrical power or synthesis gas (syngas), a combination of CO and H2. Rather than burning coal directly, gasification breaks down coal or any other organic fuel stock into basic chemical constituents. At a temperature of 800 to 1200 °C, the organic feed stock is converted to H2, CO, CO2, H2O and light gases, tars, nitrogen compounds, and solid residue through devolatilization and through reaction with the gas feed-stock of air, oxygen or steam. Gasification chemistry is generally endothermic, and heat is supplied by internal heat exchangers or by partial combustion. Gasifier configurations can be classified as fixed bed gasifiers (updraft or down draft), entrained gasifiers, and fluidized bed gasifiers. The example here is coal gasification in a fluidized bed.

CPFD governing equations

The CPFD method solves the fluid and particle momentum equations in three dimensions. The fluid is described by the Navier-Stokes equation with strong coupling with the discrete particles. The particle momentum follows the particle-in-cell (MP-PIC) numerical description from P. O'Rourke (Andrews and O'Rourke, 1996; Snider et al., 1998; Snider, 2001) which is a Lagrangian description of particle motion described by ordinary differential equations with coupling with the fluid. The CPFD solution as applied in Barracuda-CPFD® is aimed at solving commercial problems, which are generally physically large systems. In the CPFD scheme, a numerical-particle is defined where particles are grouped with the same properties (species, size, density, etc.). The numerical-particle is a numerical approximation similar to the numerical control volume where a spatial region has a single property for the fluid. Using numerical particles, large commercial systems containing billions of particles can be analyzed using millions of numerical-particles.

The equations of fluid dynamics can be derived from kinetic theory (dilute gases) or continuum points of view. The volume averaged total fluid phase mass and momentum equations (Batchelor 1988, Gidaspow 1994) are

$$\frac{\partial \theta_f \rho_f}{\partial t} + \nabla \cdot (\theta_f \rho_f \boldsymbol{u}_f) = \dot{\rho}_{sf}^{\,C} \qquad (1)$$

and

$$\frac{\partial (\theta_f \rho_f \boldsymbol{u}_f)}{\partial t} + \nabla \cdot (\theta_f \rho_f \boldsymbol{u}_f \boldsymbol{u}_f) = -\nabla p - \boldsymbol{F} + \theta_f \rho_f \boldsymbol{g} + \nabla \cdot \theta_f \boldsymbol{\tau}_f \qquad (2)$$

where $\boldsymbol{u}_f$ is the fluid velocity and θ_f is the fluid volume fraction, ρ_f is fluid density, p is fluid pressure, $\boldsymbol{\tau}_f$ is the fluid stress tensor, and $\boldsymbol{g}$ is the gravitational acceleration. $\boldsymbol{F}$ is the rate of momentum exchange per volume between the fluid and particles phases which is given below. The gas mass production rate per volume from solid-gas chemistry is $\dot{\rho}_{sf}^{C}$.

The constitutive equation for the non-hydrostatic part of the stress, τ_f, is

$$\tau_{ij}=2\mu S_{ij}-\frac{2}{3}\mu\delta_{ij}\frac{\partial u_i}{\partial u_j} \tag{3}$$

for a Newtonian fluid where μ is the coefficient of viscosity which depends only on the thermodynamic state of the fluid for laminar flow and can have an added eddy viscosity for turbulent flow. The rate of deformation S_{ij} is

$$S_{ij}=\frac{1}{2}\left(\frac{\partial u_i}{\partial x_j}+\frac{\partial u_j}{\partial x_i}\right) \tag{4}$$

The mass and momentum equations are for the total gas mixture, where the gas mixture properties are based on the mass fraction of the gases. The flow is thermal and compressible, and the gas phase pressure, temperature and density are related through an equation of state. In the CPFD scheme, an ideal gas equation of state is used where the partial pressure of gas species i is

$$p_i=\frac{\rho_i RT}{Mw_i} \tag{5}$$

and the total pressure is

$$p=\sum_{i}^{Nspecies} p_i \tag{6}$$

where R is the universal gas constant, T is mixture temperature, ρ_i is the fluid species density and Mw_i is the molecular weight of gas species i. A transport equation is solved for each gas species, and the total fluid phase properties are calculated from the mass fractions of the gas species making-up the fluid mixture. Through the breaking and combing of chemical bonds, mass is transferred between gas species which is a source or sink value in the individual gas species transport equations. For a gas-only system, the total gas phase mass, momentum and energy would remain constant. However with solid-gas reactions, mass, momentum and energy is transferred between phases and the gas conservation equations contain mass and energy source terms.

The fluid phase energy equation is

$$\frac{\partial}{\partial t}(\theta_f\rho_f h_f)+\nabla\cdot(\theta_f\rho_f h_f \boldsymbol{u}_f)=\theta_f\left(\frac{\partial p}{\partial t}+\boldsymbol{u}_f\cdot\nabla p\right)+\phi+\dot{Q}^W+\dot{Q}^C_{sf}+\dot{Q}^F_{sf}+\nabla\cdot(k_f\theta_f\nabla T_f) \tag{7}$$

where, $\dot{Q}^W$ is heat transfer from structures (walls), $\dot{Q}^C_{sf}$ is the energy transfer from solid-gas chemistry, $\dot{Q}^F_{sf}$ is the heat from the solid phase, and T_f is gas temperature. The thermal conductivity, k_f, is the molecular conductivity plus turbulent eddy conductivity. The CPFD scheme uses an ideal gas approximation for enthalpy, where the enthalpy depends only on temperature, and the enthalpy for gas species i is

$$h_i=\int_{T_{ref}}^{T} Cp_i\,dt+\Delta h_{f,i} \tag{8}$$

where C_p is the heat capacity, and Δh_f is the heat of formation of the species. The total gas enthalpy is the mass fraction weighted sum of the gas species enthalpies.

The dynamics of each numerical particle is described by a material description. The solid acceleration equation is

$$\frac{d\boldsymbol{u}_s}{dt} = D_p(\boldsymbol{u}_f - \boldsymbol{u}_s) - \frac{1}{\rho_s}\nabla p + \boldsymbol{g} - \frac{1}{\theta_s \rho_s}\nabla \tau_s \tag{9}$$

where $\boldsymbol{u}_s$ is the solid velocity, ρ_s is the solid density, θ_s is the solid volume fraction, and τ_s is the solid normal stress. The equation for solid movement is

$$\frac{d\boldsymbol{x}_s}{dt} = \boldsymbol{u}_s \ . \tag{10}$$

The solid energy equation is

$$C_v \frac{dT_s}{dt} = \frac{A_s}{m_s}\gamma_s (T_f - T_s) + \frac{1}{m_s}\dot{h}^C_{fs} \tag{11}$$

where T_s is the solid temperature, γ_s is the heat transfer coefficient between solid and gas, and $\dot{h}^C_{fs}$ is the energy transfer from gas-solid chemistry. Conservation of mass requires the solid mass to change based on gas-solid chemistry, and the mass loss or gain from solids has a corresponding loss or gain in the gas phase. Because a solid density is constant, the change in solid mass is a reduction in particle size.

In the MP-PIC scheme, solids properties are mapped to and from the Eulerian grid to get grid-properties for the solids. Fluid properties, are in turn, mapped to discrete particle locations. The interpolation operator is the product of interpolation operators in the three orthogonal directions. For a particle located at $\boldsymbol{x}_p$, where $\boldsymbol{x}_p = (x_p, y_p, z_p)$, the x-directional component of the interpolation operator to grid cell i, is an even function, independent of the y and z coordinates, and has the properties.

$$S_i^x(x_p) = \begin{matrix} 0 & x_{i-1} \geq x_p \geq x_{i+1} \\ 1 & x_p = x_i \end{matrix} \ . \tag{12}$$

The y and z interpolation operators have a similar form. The particle volume fraction at cell ξ from mapping particle volume to the grid is

$$\theta_{s\xi} = \frac{1}{V_\xi}\sum_{t}^{N_s} V_s n_s S_{s\xi} \tag{13}$$

where V_ξ is the grid volume, V_s is particle volume, n_s is the number of particles in a numerical particle, and the summation is over all numerical particles N_s. From conservation of volume, the sum of fluid and particle volume fractions equals unity, $\theta_s + \theta_f = 1$.

The interphase drag coefficient is

$$D_s = C_d \frac{3}{8}\frac{\rho_g}{\rho_s}\frac{|\boldsymbol{u}_f - \boldsymbol{u}_s|}{r_s} \tag{14}$$

and the Wen and Yu (1966) drag correlation for C_d is used in this study:

$$\begin{aligned} &\mathrm{Re} < 1000 \quad C_d = \frac{24}{\mathrm{Re}}\left(1 + 0.15\,\mathrm{Re}^{0.687}\right)\theta_f^{-2.65} \\ &\mathrm{Re} \geq 1000 \quad C_d = 0.44\,\theta_f^{-2.65} \end{aligned} \tag{15}$$

$$\mathrm{Re} = \frac{2\rho_f |\boldsymbol{u}_f - \boldsymbol{u}_s| r_s}{\mu_f} \ , \tag{16}$$

where μ_f is molecular the gas viscosity and the particle radius is

$$r_s = \left(\frac{3V_s}{4\pi}\right)^{1/3} \tag{17}$$

The particle normal stress is an approximation of collective effects of neighbor particles on a particle. The MP-PIC method (Andrews and O'Rourke, 1996) makes use of spatial gradients which are readily calculated on the Eulerian grid and apply the gradient to discrete particles. Gradients in this particle pressure result in particle accelerations that prevent particle volume fractions from exceeding their close-pack limit. The particle pressure is a monotonic increasing function of solids volume fraction and vanishes as the solids volume fraction goes to zero. Particle-to-particle collisions are modeled by the particle normal stress. The particle stress is derived from the particle volume fraction which is in turn calculated from particle volume mapped to the grid. The particle normal stress model used here is from Harris and Crighton (1994)

$$\tau = \frac{P_s \theta_s^{\beta}}{max\left[\left(\theta_{CP}-\theta_s\right), \epsilon\left(1-\theta_s\right)\right]} \quad , \tag{18}$$

where ε is a small number on the order of 10^{-7} to remove the singularity. Snider (2001) applied the solid normal stress model with a dependency on each solid's velocity vector. Solid collisions depend on solids concentration and the deviation of solids velocity from a solids equilibrium velocity. The model applies the particle normal stress to a solid up to the point where the solid reaches the particle-mean velocity. The model further assumes that near close pack, the multiple collisions occurring over the hydrodynamic calculation interval will move solids to relieve high stress, but the multiple collisions limit motion to near the particle-mixture mean velocity. Therefore particle collisions near close pack will not impart a solids velocity greater than the mean particle-mixture velocity.

The fluid momentum equation implicitly couples fluids and particles through the interphase momentum transfer. The interphase momentum transfer at momentum cell ξ is

$$\boldsymbol{F}_{\xi}^{n+1} = \frac{1}{V_{\xi}} \sum_{k}^{N_p} S_{\xi,sk} \left[D_{sk}\left(\boldsymbol{u}_{f,sk}^{n+1} - \boldsymbol{u}_{sk}^{n+1}\right) - \frac{1}{\rho_{sk}} \nabla p_{sk}^{n+1} \right] n_{sk} m_{sk} \quad , \tag{19}$$

where m_s is the solids mass, n_s is the number of solids in a numerical particle, S is the interpolation operator and the subscript "s" on fluid properties (p, $\boldsymbol{u}_f$) indicate the fluid property is interpolated to a particle location. The summation is over all solids. The heat transfer from the solids phase to gas phase for cell ξ is is

$$\dot{Q}_{\xi,sf}^{F} = \sum_{\kappa}^{N_p} S_{\kappa\xi} m_{s\kappa} \left[\dot{h}_{fs\kappa} + \dot{h}_{fs\kappa}^{C} - D_{s\kappa}(u_{f\kappa} - u_{s\kappa})^2 \right] \tag{20}$$

where $\dot{h}_{fs}$ is the convective solid-gas heat transfer, $\dot{h}_{fs}^{C}$ is the energy from gas-solid chemistry, and the last term on the right is energy from solid-gas drag.

The CPFD scheme approximates conservation equations by finite volumes with staggered scalar and momentum nodes. A large eddy simulation (LES) turbulence model is used where large eddies are directly calculated and the subgrid-scale turbulence is modeled (Smagorinsky,1963). The Smagorinsky model is an eddy viscosity model which is based on the notion that the effects of subgrid-scale Reynolds stress increases transport and dissipation. The mixture fluid density, velocity and pressure are coupled by a semi-implicit pressure equation derived from the gas mass conservation equation which is applicable to an arbitrary Mach number (Ferziger and Peric 1996). The momentum, energy and pressure equations are solved with a conjugate gradient solver, and each particle momentum, motion and energy ordinary differential equations are directly solved.

There are thousands of reactions in a gasifier or other industrial chemical processes. However, even if all the elemental reactions and their rates of reaction could be identified, it is not possible to calculate this large number of coupled reactions for a large reactor over 100s of seconds simulation time. The approach is to postulate a reduced set of reactions which describe the major reactions in the reactor. The rate coefficients for a postulated set of reactions are empirically fitted with measured data.

In the CPFD method, the chemistry can defined in a number of formats. In one format, the rate of reaction of a gas species is specified directly as a set of ordinary differential equations which are solved implicitly, such as

$$\begin{aligned} R_{VGO} &= -(k_1+k_2+k_4)C_{VGO}^2 - k_3 C_{VGO}^2 \\ R_{Gasoline} &= \left[k_1 C_{VGO}^2 - (k_5+k_7)C_{Gasoline}\right] - k_6 C_{Gasoline} \\ R_{HGas} &= (k_2 C_{VGO}^2 + k_5 C_{Gasoline}) \\ R_{LGas} &= k_3 C_{VGO}^2 + k_6 C_{Gasoline} \\ &\vdots \quad \vdots \end{aligned} \tag{21}$$

where the rate for gas i is $R_i = dC_i/dt$, and C is a mass or mole concentration. In the example, the subscripts are for the complex gas groups of vacuum gas oil (*VGO*), gasoline (*Gasoline*), heavy gas (*HGas*) and light gas (*LGas*). The affect of a catalyst is contained within the rate coefficient which is of the form

$$k_i = a_i \theta_{s,k} exp\left(\frac{-E_i}{T}\right) \tag{22}$$

where a_i is a constant and $\theta_{s,k}$ is the volume concentration (volume fraction) of catalyst k. Each of the gas species has its own transport equation. The input of rate equations is straight forward through a graphical interface and there is flexibility in the form of the rate coefficient. The set of ordinary differential equations (ODE) are solved implicitly at each spatial cell and each time step. Because the chemistry shown above is catalytic, there is no energy or mass source in the fluid equations (no solids chemistry).

Another method of specifying chemistry in the CPFD method is by specifying stoichiometric type chemistry and a rate for each of the stoichiometric equations, such as

$$\begin{aligned} &F + 2B \rightarrow 3E + D \\ &C(S) + B \rightarrow 2A + H \\ &\vdots \qquad \vdots \end{aligned} \tag{23}$$

The solid material is specified as *C(S)*. The rate for the 1st stoichiometric equation is

$$R_1 = k_1 [F][B]^2 \tag{24}$$

and has a typical rate coefficient of

$$k_1 = T^3 A_{s,C} e^{-E_1/T + E_0} \tag{25}$$

Each stoichiometric equation has a rate equation. The *[F]* specifies the mole concentration of material *F*. The rate coefficient, shown above, depends on temperature, *T*, and the solid surface area of species *C* per volume. Again, the input of rate equations is straight forward through a graphical interface and there is flexibility in the form of the rate coefficient. The set of stoichiometric and rate equations are automatically put in the form of a set of ordinary differential equations (ODE) and solved implicitly at each spatial cell and each time step. The change of gas species plus any mass source from solid chemistry is included within the transport equation for each gas species.

The solid properties are mapped to the grid and the grid-solid properties are used in rate equations. In the example above the solid surface area per volume is the sum of surface areas of all solids in the cell divided by cell volume. Using average solid properties per cell such as Sauter mean diameter, solids surface area, etc., is justified based on the accuracy of reduced chemistry equations.

The advantage of using an average solid property is that the solids are semi-implicitly coupled in the solution of the rate ODE equations. After solution of chemistry, the mass change of each discrete solid is increased or decreased based on the total solid mass change in the cell. While the CPFD method has options to explicitly solve the chemistry for each discrete solid, the method can be computationally expensive.

In this work, a "shrinking" solid model is used. That is, the solid material-density is assumed constant and the solid size changes as it gains or looses mass. In this study, the solids are one material such as carbon or ash, but a solid is not a combination of materials.

Coal gasifier example

The example of CPFD thermal-chemistry is a generic coal gasifier shown in Fig. 1. The gasifier is a large 4.57 m (15 ft) diameter and 13.7 m (45 ft) high cylinder with a dome top. The reactor has four internal cyclones, and because the chemistry is endothermic, a thermal boundary condition is specified for several internal heating tubes and on the walls in the bottom portion of the reactor to provide heat. Steam and oxygen are fed from a sparger at the bottom of the reactor. Gas and entrained solids exit through the four cyclones. Each cyclone dipleg is configured to feed particles into the system at the same rate as solids are lost into the cyclone (100% efficient cyclones) using the Barracuda controller. In an operating unit, the cyclones would lose some portion of the entrained particles, and there would be a fresh coal feed to replenish the particles. For the 600 s simulation time, it is acceptable to neglect cyclone losses and coal feed and simply keep the system mass constant. The gasifier operating conditions are given in Table 3. Initially, the reactor is filled completely with N_2. About 47,000 kg (104,000 lbs) of solids are initially in the vessel.

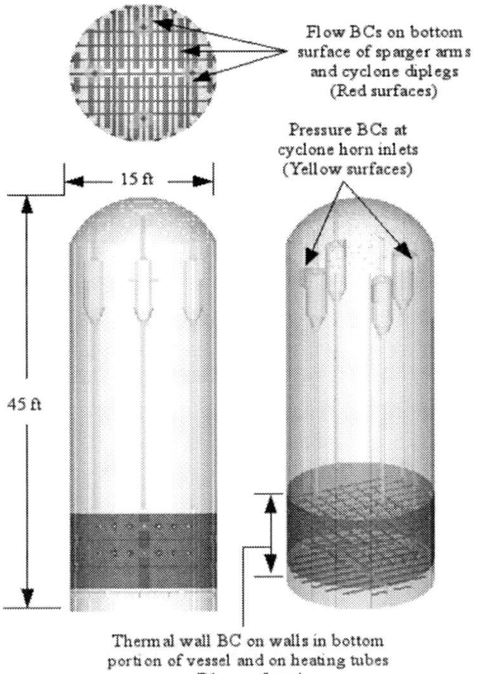

Figure 1: Gasifier geometry and boundary conditions

The composition of coal can be categorized as moisture, volatiles, tar, ash and carbon. At gasifier temperature of 800 to 1200 °C, vaporization and volatization occurs quickly. Coal within the bed has lost its moisture and volatile gases. Ash does not take part in any reactions. The carbon reacts with the feed stock of oxygen and steam and the chemical reactions are given in Table 1.

Table 1. The gasifier reduced-set stoichiometric equations.

Steam gasification	$C(s) + H_2O \leftrightarrow CO + H_2$
CO_2 gasification	$C(s) + CO_2 \leftrightarrow 2CO$
Methanation	$C(s) + 2H_2 \leftrightarrow CH_4$
Combustion	$\lambda C(s) + O_2 \rightarrow 2(\lambda\text{-}1)CO + (2\text{-}\lambda)CO_2$
Water gas shift	$CO + H_2O \leftrightarrow CO_2 + H_2$

The rates for the stoichiometric equations are given in Table 2. The chemistry rates are obtained from a number of sources in the open literature. The gasification chemistry is from Wen (1982), as used by Syamlal (1992). The kinetic rate expression for the combustion reaction is from Yoon (1978), and the value of λ is assumed to be 2, such that only CO is produced from carbon combustion. The water gas shift is adopted from Bustamante (2004, 2005). Because reduced chemistry is a postulated set of reactions, and the rate coefficients should be fitted as a set to measured data, it is not recommended to mix rates from different sources. It is not reasonable to expect an accurate calculation of chemical reactions by picking and choosing the reaction equations and rates from various sources. However. because there is not a complete, consistent set of gasifier chemistry equations and rates in the open literature, there is no choice but to use unrelated chemistry rates.

The chemistry is described by five reactions (nine reactions considering reverse reactions) given in Table 1. The reaction rate is in *gmol/m³s*, solids mass per volume, m_s, is in *kg/m³*, temperature, *T*, is in *K*, and *[]* implies mole concentration (*gmol/m³*) of the gas species enclosed in the brackets. Both forward and reverse reaction rates are calculated in lieu of a combined forward-reverse rate. The motivation for separation of the reverse and forward reactions is to determine dominant reactions in the dynamic gasifier. The solid is modeled as pure carbon with a particle size distribution shown in Fig. 2. For the gasifier solid-gas chemistry, it may have been more appropriate to define (fit) the chemical rates based on the solids surface area; however the authors of the chemistry rate equations chose solid mass per volume. The solid-gas chemistry consumes the solids and the solids shrink from chemistry.

The CPFD provides a number of methods of specifying chemistry. In this study, the chemistry is specified as stoichiometric chemical reactions with rates as described above. The chemistry is entered in Barracuda much as shown in Tables 1 and 2. The CPFD method formats the equations into solvable ordinary differential equations (ODEs) and builds the Jacobian for the solution of ODE set at each time and location. The equations are often sparse and stiff and require a robust implicit solver (Hindermarsh and Serban, 2006). In this study, reactions were spit into reverse and forward reactions giving nine ODE to solve.

Table 2: Chemical equations and reaction rates

Reaction	Stoichiometric Equation	Reaction Rate Expression (mol m^{-3} s^{-1})	Exothermic / Endothermic
Steam gasification	$C(s)+H_2O \rightarrow CO+H_2$	$r_{1,f}=1.272 m_s T \exp\left(\frac{-22,645}{T}\right)[H_2O]$	Exothermic
	$CO+H_2 \rightarrow C(s)+H_2O$	$r_{1,r}=1.044\times10^{-4} m_s T^2 \exp\left(\frac{-6,319}{T}-17.29\right)[H_2][CO]$	Endothermic
CO_2 gasification	$C(s)+CO_2 \rightarrow 2CO$	$r_{2,f}=1.272 m_s T \exp\left(\frac{-22,645}{T}\right)[CO_2]$	Exothermic
	$2CO \rightarrow C(s)+CO_2$	$r_{2,r}=1.044\times10^{-4} m_s T^2 \exp\left(\frac{-2,363}{T}-20.92\right)[CO]^2$	Endothermic
Methanation	$0.5C(s)+H_2 \rightarrow 0.5CH_4$	$r_{3,f}=1.368\times10^{-3} m_s T \exp\left(\frac{-8,078}{T}-7.087\right)[H_2]$	Exothermic
	$0.5CH_4 \rightarrow 0.5C(s)+H_2$	$r_{3,r}=0.151\, m_s T^{0.5} \exp\left(\frac{-13,578}{T}-0.372\right)[CH_4]^{0.5}$	Endothermic
Combustion	$2C(s)+O_2 \rightarrow 2CO$	$r_4=4.34\times10^7 \theta_p T \exp\left(\frac{-13,590}{T}\right)[O_2]$	Exothermic
Water gas-shift	$CO+H_2O \rightarrow CO_2+H_2$	$r_{5,f}=7.68\times10^{10} \exp\left(\frac{-36,640}{T}\right)[CO]^{0.5}[H_2O]$	Exothermic
	$CO_2+H_2 \rightarrow CO+H_2O$	$r_{5,r}=6.4\times10^9 \exp\left(\frac{-39,260}{T}\right)[H_2]^{0.5}[CO_2]$	Endothermic

Table 3. Gasifier operating conditions

Superficial gas velocity	0.4 m/s (1.31 ft/s)
Sparger inlet gas composition by mass	80% H_2O, 20% O_2
Sparger inlet gas temperature	1,200 K (1,700 °F)
Sparger inlet gas superficial velocity	1.0 m/s (3.28 ft/s)
Static bed height	4.6 m (15 ft)
Bed mass	47,000 kg (104,000 lbs)
Solids close pack volume fraction	0.55
Initial gas in reactor	100% N_2
Initial solid and gas temperature	1,200 K (1,700 °F)
Initial gasifier pressure	1720 kPa (250 psig)

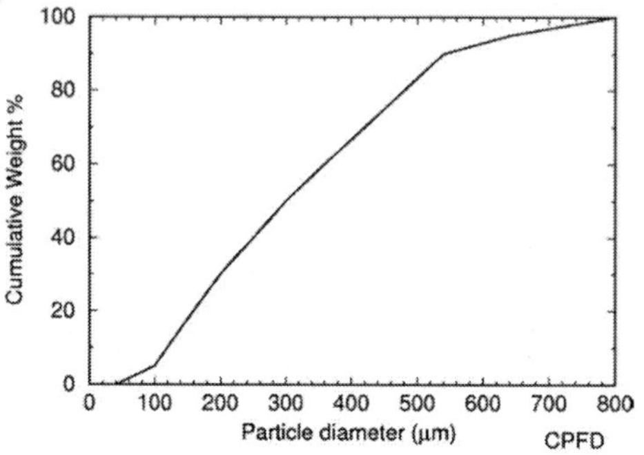

Figure 2: Particle size distribution

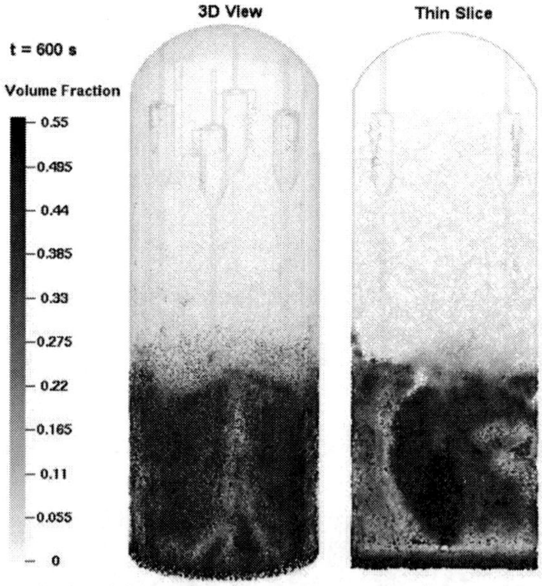

Figure 3: Particles colored by volume fraction at t = 600 s

The CPFD method has been applied and validated for a variety of fluidized beds [Snider, D. and Banerjee, S., 2009]. In this application the geometry is relatively complex with internal cyclones and gas spargers. The particle size distribution used for this simulation, as shown in Fig. 2, is classified as a Geldart group B material. The fluidization behavior observed is typical of what one would expect for a group B material at a superficial gas velocity of 0.4 m/s (1.3 ft/s). Specifically, the bed is in a churned-turbulent state and is well-mixed. There are distinct regions of high and low volume fraction, and large void structures ("bubbles") are observed in the bed.

Fig. 3 shows a full view of the reactor and a view with the front half of the reactor cut-away with particles colored by volume fraction. Darker color indicates higher solids concentration (higher solids volume fraction). Fig. 4 shows calculated parameters with the front of the vessel removed. The freeboard region is quite dilute, and there is a distinct transition from the dense particle bed to the dilute solids in the freeboard. Generally a fluidized bed without internal structures is relatively well mixed. With internal structures, the gas-solid flow pattern is affected by the structures. An interesting example of the affect of a structure is the the CPFD predicted effect the sparger center-delivery-pipe has on the local solids volume fraction. The pipe delivers gas to the sparger arms but does not directly discharge gas into the bed. The cut-image in Fig. 3 (also seen in Fig. 4) shows the end of the sparger delivery-pipe and the high concentration of solids which forms above the pipe. The delivery pipe blocks fluidizing gas and initiates the dense region, and subsequently there is a persistent tendency for fluidizing gas to flow around this packed region rather than through it. The dynamics and mixing of solids in the bed will affect the overall performance of the gasifier.

The CPFD method as discussed above describes the energy on an enthalpy basis. In this example heat is supplied from walls and heating tubes, and energy is added with the gas feed. The breaking and forming of chemical bonds changes the bed temperature, and solids shrink from the CPFD solid-gas chemistry. The gasifier provides a complex application of the CPFD thermal model with chemistry. The calculation begins with a uniform gas and solids temperature of 1200 K. The gas feed at the sparger introduces 80% H_2O and 20% O_2, by mass, at a temperature of 1200 K. Oxygen is supplied in the gas feed to give thermal heat from combustion to overcome the endothermic gasification reactions.

One advantage of using a fluidized bed for any reactor is that it tends to maintain a uniform temperature throughout the bed. The mixing of solids and the large solids surface area gives a relatively uniform bed temperature, and the large thermal mass of solids gives a slow changing bed bulk temperature. Fig. 4c and Fig. 5 shows the gasifier is thermally well mixed in the central region of the bed with a gas temperature ranging from 1125 to 1200 K. Near the heated walls the temperature is higher as seen in Fig. 4c. The region below the sparger is at 750 K, which is significantly lower than the bulk bed temperature. The cool region below the sparger is from a number of combined effects. First, the solids below the sparger are not fluidized and heat transfer from bed mixing does not reach the stagnant layer below the sparger. Second, the wall heating is above the sparger and there is no external heat applied below the sparger. Third, the hot combustion of gas produced at the sparger moves up and mixes in the bed. Fourth, the cool region below the sparger is conducive to the reverse reaction of CO_2 gasification ($2CO \rightarrow C(s)+CO_2$), which is endothermic.

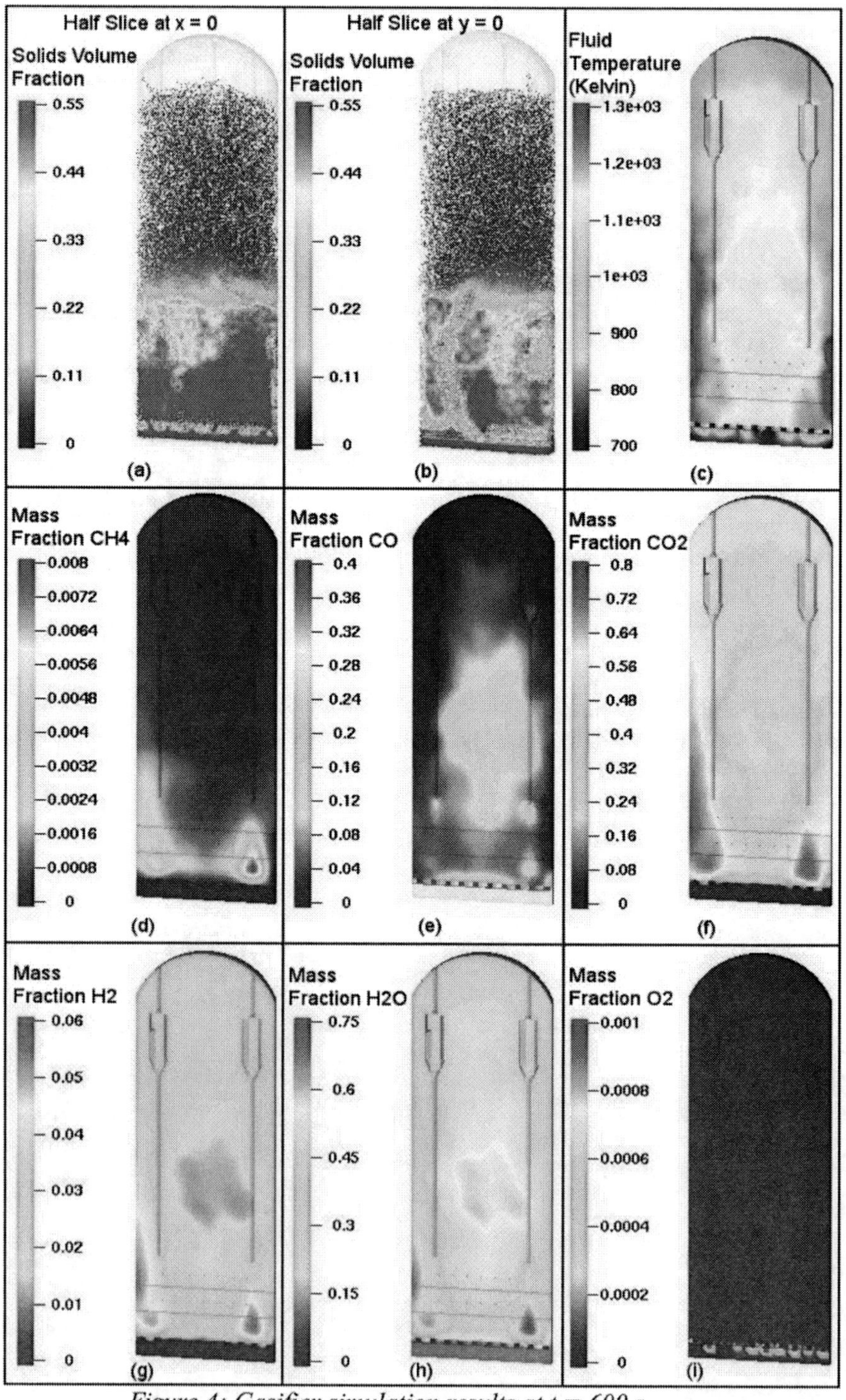

Figure 4: Gasifier simulation results at t = 600 s

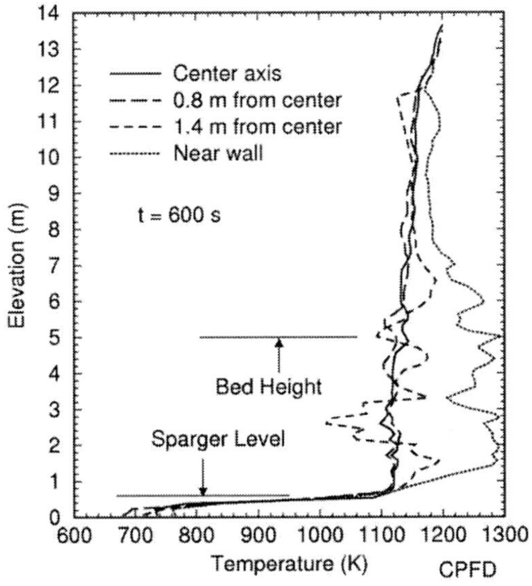

Figure 5: Temperature along the gasifier central axis at t = 600s.

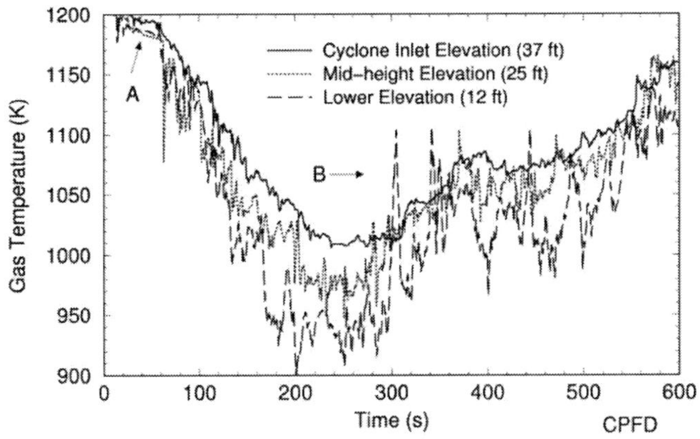

Figure 6: Gas temperature at three elevations vs time

The CPFD method includes a robust representation of chemistry as described above. The chemistry affects the bed dynamics by locally changing the gas mixture properties and by changing the bed temperature from breaking and forming chemical bonds. The chemical-energy to thermal-energy is handled well by enthalpy-base energy equation. The CPFD method calculates chemical reaction rates at every computational time-step for each computational cell in the simulation. This enables an extremely detailed analysis of reaction kinetics and gas species concentrations coupled with bed dynamics throughout the calculation domain, and provides a wealth of data for studying the gasifier.

Fig. 7 shows the chemical reaction rates vertically up the central region of the gasifier. The combustion reaction is fast and oxygen introduced through the sparger is instantly consumed, as shown in Fig. 8. Gasification also occurs directly above the sparger with the injection of hot steam. Fig. 8 shows the gasification by the decreases in H_2O from reaction with solid carbon, and the increases in H_2

and CO_2 above the sparger. The H_2 and CO_2 are from the water gas-shift following gasification, which occurs almost instantly. The chemistry involving the solid carbon reduces or increases the size of the solid. The change in carbon mass, as discussed above, requires a mass and energy transfer between the gas and solid phases. Moving up in the fluidized bed, the forward rates of steam gasification, CO_2 gasification, methanation, and water gas-shift become dominant. CO_2 reacts with carbon through CO_2 gasification, while H_2 is consumed through methanation and reverse water gas-shift. The solids in the bed mix with upward and downward motion. The gas, on the other hand, moves up through the bed with little downward mixing. In the freeboard region above the fluidized bed, the solids concentration is low and the water gas-shift reaction, which is independent of solids concentration, is the dominant reaction. The near equilibrium balance achieved by the fast water gas-shift determines the final concentration of gas species at the cyclone inlets.

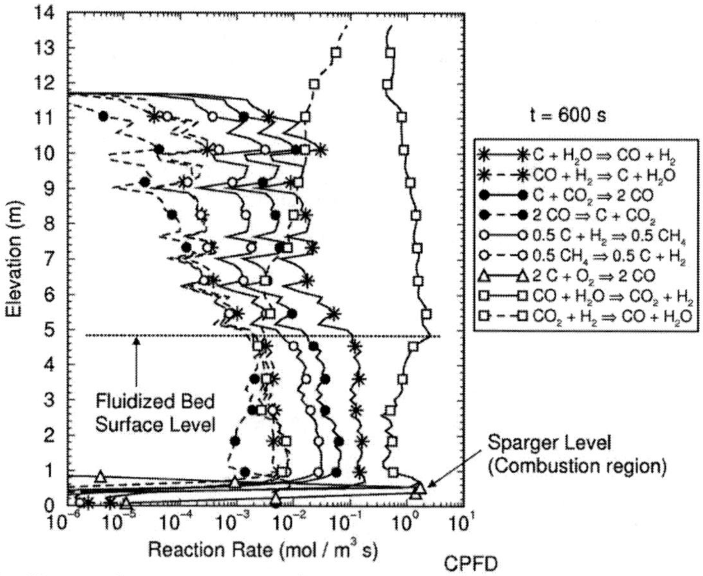

Figure 7: Chemical reaction rates along central axis of reactor at t = 600 s

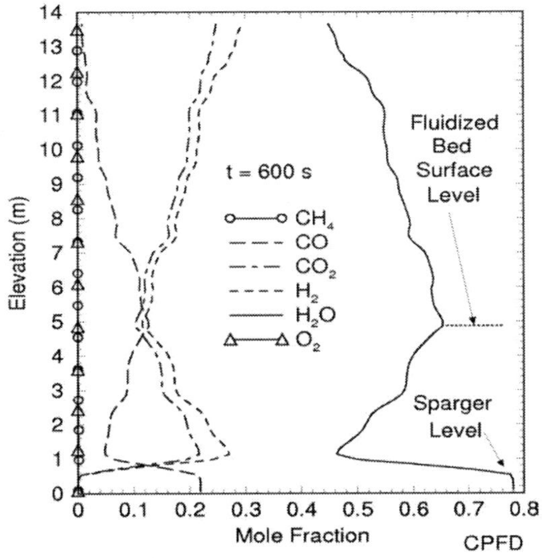

Figure 8: Gas species mole fractions along central axis of reactor at t = 600 s

Because the gasification chemistry is endothermic, oxygen was supplied to increase the thermal energy and external heat was supplied through the vessel wall and from internal heating tubes. The target bed temperature is 1200 K. The gasification chemistry used in this study is highly endothermic and it was found that more heat than originally estimated was needed to keep the target temperature. Fig. 6 shows the transient bed temperature from start to 600 s. While the calculated transient data leading to a steady condition is usually not useful because it does not match plant start up, in this case, it illustrates the difficulty in supplying sufficient external heat to maintain the target temperature. For the first 60 s (point *A* in Fig. 6), the thermal mass of the bed was sufficient to maintain the bed temperature. However, by 200 s (point *B* in Fig. 6), it became apparent that more heat was needed to keep the bed at the specified operating temperature. The heat transfer was increased, and by 600 s the reactor was at the target temperature. At 600 s, the difference between gas temperatures at 3.7 m (12 ft) and 11.3 m (37 ft) elevations is about 50 K, giving a nice uniform bed temperature.

The reactions are temperature dependent and the effect of temperature can be seen in the images in Fig. 4. The temperature next to the heated wall is higher than in the central core of the gasifier (Fig 4c), and the high temperature promotes gasification as seen by the high concentration of CO_2 and reduced H_2O concentration. CH_4 is also produced directly above the sparger and along the wall where the gas-solids are at high temperature and there is a rich environment of H_2. The cold region below the sparger consists of primarily CO and H_2O.

The gas composition obtained at the cyclone inlets is an important metrics of a gasification system. The purpose of the gasifier is to generate valuable product gases. Fig. 9 shows the mole fractions of each gas species at the cyclone inlet horns over the course of the simulation. The change in gas species shown in Fig. 9 along with the gas temperature shown in Fig. 6, shows the product gas composition depends on the bed temperature. When the gas temperature in the reactor reached a low point around 200 s, the concentrations of H_2O and CO reached a corresponding high point due to reduced gasification and a water gas-shift favoring the reverse direction. Similarly, as the gasifier's temperature returned to the 1150 K range, the concentrations of H_2 and CO_2 also rebounded. The water gas-shift reaction reaches equilibrium quickly, and the balance between forward and reverse reactions achieved is dependent on temperature.

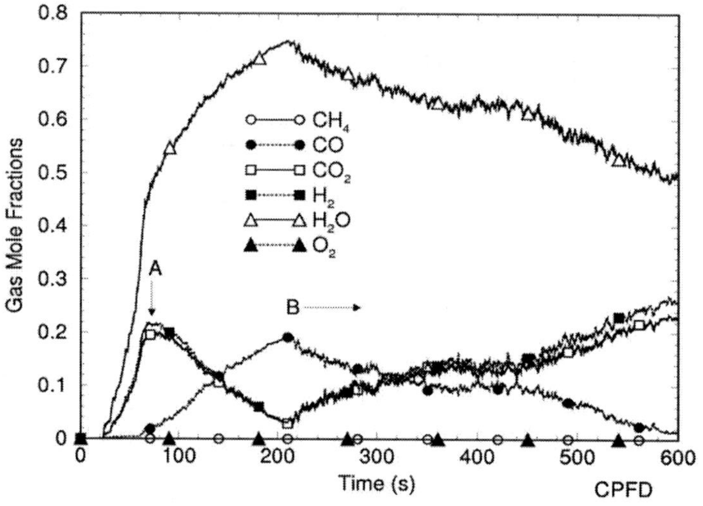

Figure 9: Product gas species mole fractions at cyclone inlets at t = 600 s

The chemical reactions and kinetic rates used in this study, at the target temperature of 1200 K, lead to an outlet gas stream with relatively high concentrations of H_2O, H_2, and CO_2. This combination of product gases contains H_2, which is valuable, as well as CO_2 which is undesirable. If CH_4 or a more standard syngas product with a higher concentration of CO were desired, the design and operating conditions of the reactor would need to be altered.

Another metric for the performance of a gasifier is the efficiency of the overall process. The "hot gas efficiency" (thermal efficiency) is used in situations where the gas is not cooled before combustion, and is defined as (Basu, 2006)

$$\eta_{\text{geff}} = \frac{V_g q_g + H_{\text{sensible}}}{M_b C_b} \tag{26}$$

where V_g is the gas generation rate, q_g is the heating value of gas, $H_{sensible}=C_p V_g \rho_g (T_g - T_a)$, C_p is the specific heat of gas, ρ_g is the gas density, T_g is the gas temperature, T_a is the ambient temperature (or feed temperature of inlet gas, if not ambient), M_b is the fuel consumption rate, and C_b is the heating value of fuel. For the gasifier in this study, using typical heating values of coal and product gases (Basu 2006, Swanson, 1976), the efficiency is $\eta_{geff} = 0.68$. This is a reasonable efficiency for the generic gasifier in this study.

Conclusions

Thermal and chemistry in the CPFD Eulerian Lagrangian method is described. The thermal and chemistry equations are solved in a coupled solution with the fluid-solid flow. The chemistry is specified in a general format which allows flexibility, and the chemistry can be gas-gas homogenous chemistry, gas-solid catalytic chemistry, gas-solid heterogeneous chemistry or any combination of the types of chemistry. There is virtually an unlimited number of reactions, gas species and solids. The gas species each have their own transport equation, and the gas mass may change from solid chemistry. The solids Lagrangian motion is described by the MP-PIC method and each numerical solid has its own momentum and energy equation. The solid sizes may change from solid chemistry. The energy equation is formulated in terms of enthalpy. This provides a direct calculation of chemical to thermal energy as the gas species change through breaking or forming chemical bonds.

The CPFD method is applied to a syngas gasifier. The gasifier is a complex reactor with internal cyclones, internal heaters and gas sparger. While the gasifier in this study is generic, it illustrates the ability of the CPFD method to calculate complex gas solids flow with chemistry in a large three-dimensional reactor. The CPFD method provided a smooth solution to 600 s. The simulation time could have been shorter, but the external heat source required major adjustment (increased heat) during the simulation to keep the bed target temperature of 1200 K.

While the purpose of the problem is to illustrate the CPFD method for solving energy and chemistry, the analysis gave a wealth of information on a generic gasifier behavior, including fluidization, thermal, and chemical characteristics. While the open literature chemistry leaves room for improvement, the reduced chemistry and reaction rates provide a good starting point for analyzing commercial coal gasifiers. Further, the generic geometry contains sufficient detail such as cyclones and gas sparger, illustrating the complexity of problem which can be solved by the CPFD method.

References

Andrews, M. J. and P. J. O'Rourke, "The multiphase particle-in-cell (MP-PIC) method for dense particle flow," *Int. J. Multiphase Flow*, **22**, 379-402 (1996).

Amsden, A. A., O'Rourke, P. J., and Butler, T. D., 1989, "*KIVA-II:* A computer program for chemically reactive flows with sprays," LA-11560-MS, Los Alamos National Lab., Los Alamos, NM.

Basu, P., "Combustion and gasification in fluidized beds", CRC Press, 2006.

Batchelor, G. K., "A new theory of the instability of a uniform fluidized bed," *J. Fluid Mech.*, **193**, 75-110 (1988).

Bustamante, et al., "Uncatalyzed and Wall-Catalyzed Forward Water-Gas Shift Reaction Kinetics", presented in AIChE Journal, March 2005.

Bustamante, et al., "Kinetics of the Homogeneous Reverse Water-Gas Shift Reaction at High Temperature", presented in AIChE Journal, April 2004.

Cundall, P. A. and Strack, O. D. L. "The discrete numerical model for granular assemblies," *Geotechnique,* **29**, 1,47-65,1979.

Ferziger, J.H. and Peric, M., *Computational methods for fluid dynamics,*" Springer-Verlag Berlin Heidelberg New York, 1996.

Gidaspow, D., "Hydrodynamics of fluidization and heat transfer supercomputer modeling," *Appl. Mech. Rev.* **39**, 1-22 (1986)

Gidaspow, D., 1994, *Multiphase flow and fluidization continuum and kinetic theory description*, Academic Press, Boston.

Godlieb, W., Deen, N. G. and Kuipers J.A.M., "A discrete particle simulation study of solids mixing in a pressurized fluidized bed," *2007 ECI Conference on the 12th International Conference on Fluidization -New Horizons in Fluidization Engineering, Vancouver, Canada.*

Hirshfelder, J.O., Curtiss, C.F, and Bird, R.B., *Molecular Theory of Gases and Liquids*, John Wiley & Sons, NY, NY, 1954.

Jackson, R, *The dynamics of fluidized particles*, Cambridgw University Press, Cambridge, UK, 2000.

Karri, S.B., Issangya, A. and Knowlton, "Gas Bypassing in deep fluidized beds," *Fluidization XI,* eds. U. Arena, R. Chirone, M. Miccio and P. Salatino, Engineering Foundation, New York, 515-521, 2004.

O'Rourke, P. J., *Collective drop effects on vaporizing liquid sprays*, Ph.D. Thesis, Princeton University, 1981.

Snider, D. M., P. J. O'Rourke, and M. J. Andrews, "Sediment flow in inclined vessels calculated using multiphase particle-in-cell model for dense particle flow," *Int. J. Multiphase Flow*, **24**, 1359-1282 (1998).

Snider, D. M., "An Incompressible three dimensional multiphase particle-in-cell model for dense particle flows", *Journal of Computational Physics*, **170**, 523-549 (2001).

Snider, D. and Banerjee, S., 2009. "Heterogeneous gas chemistry in the CPFD Eulerian–Lagrangian numerical scheme (ozone decomposition)", Powder Technology. http://dx.doi.org/10.1016/j.powtec.2009.04.023

Smagorinsky, J., General circulation experiments with the primitive equations, pat I: the basic experiment," *Monthly Weather Review*, **91**, 99-164, 1963.

Swanson, V. E., "Collection, chemical analysis, and evaluation of coal samples in 1975", Open-file report 76-468, U.S. Geological Survey, 1976. Available for download at: http://pubs.er.usgs.gov/djvu/OFR/1976/ofr_76_468.djvu

Syamlal, M., and Bissett, L.A., 1992, *METC Gasifier Advanced Simulation (MGAS) Model*, DOE/METC--92/4108, DE92 001111.

Wen, C.Y. and Yu, Y.H., (1966) "Mechanics of fluidization," Chem. Eng. Progr. Symp., Ser. 62, 100-110.

Wen, C.Y., Chen, H., and Onozaki, M., 1982, *User's Manual for Computer Simulation and Design of the Moving Bed Coal Gasifier*, DOE/MC/16474-1390, NTIS/DE83009533.

Yoon, H., Wei, J., and Denn, M., 1978, "A Model for Moving-Bed Coal Gasification Reactors," *AIChE Journal,* **24**, 5.

Coarse Discrete Element Simulations for Gas-Solids Flows

Sreekanth Pannala, Computer Science and Mathematics Division, Oak Ridge National Laboratories, Oak Ridge, TN

Janine E. Galvin, Chemical and Biological Engineering, University of Colorado, Boulder, CO

Sofiane Benyahia, National Energy Technology Laboratory, Department of Energy, Morgantown, WV

Rahul Garg, Mechanical Engineering, Iowa State University, Ames, IA

Chris Guenther, National Energy Technology Laboratory, U.S. Department of Energy, Morgantown, WV

Discrete element model (DEM) coupled with the gas-phase continuum solvers are computationally very expensive to perform simulations of realistic devices. In this paper, we present an alternative approach inspired by the recent success of MP-PIC (Multiphase Particle-in-cell) methods to improve the computational efficiency by conducting DEM on particle clouds rather than individual particles. There are many open questions about strategies for good approximations for the coarse-grained systems and we will be exploring and evaluating some of those possibilities in this paper along with comparing the methods with existing continuum simulation capability.

In the coarse DEM simulation, each particle represents N number of original particles at close packing and these coarse particles are subjected to collisions etc. using soft-sphere spring-dashpot DEM model and the drag computations are based on the original particle size and density. The major assumption into this model is that the

internal collisions within the parcel do not have any first-order effects and the parcels are homogeneous. The pros of this approach are:

- Drastically reduce the computational costs

- Includes both normal and tangential forces

- Address close-packing naturally

- Account for particle rotation

- Superior solids advection through the Lagrangian tracking of the particles

- Use existing Eulerian-Lagrangian Framework

- Can be extended easily to multiple particle sizes

Some of the cons of this approach are:

- No mathematical framework to bound the error from the approximation

- There is no way A priori to judge the sampling accuracy and efficiency

In this paper we present what-if numerical experiments of a fluidized bed with a central jet using our Coarse DEM model. We find that the these simulations are able to capture the two dominant modes in the power spectrum density of pressure while the bed height variation seems to have the correct qualitative behavior. Comparing these results to comparable continuum-continuum simulations shows that the ability to convect the particles accurately might have a big influence on the ability of continuum-discrete to have better predictability. We will report on the role of the parameter space (in terms of numerical approximations) on the differences between the various simulations and make recommendations on the acceptable coarsening strategies. We hope that this work would lay down some groundwork for more rigorous coarsening strategies for discrete-continuum simulations for gas-solids flows and improve drastically the efficiency of current techniques.

See more of this Session: Computational and Numerical Approaches to Particle Flow II
See more of this Group/Topical: Topical 4: Computational Particle Technology

An Exact Method for Determining Local Solid Fractions in Discrete Element Method Simulations

Ben Freireich, School of Mechanical Engineering, Purdue University, West Lafayette, IN

Madhusudhan Kodam, Mechanical Engineering, Purdue University, West Lafayette, IN

Carl R. Wassgren, Purdue University, West Lafayette, IN

A solid fraction algorithm is presented which accounts for the partial volume of a sphere straddling cuboidal measurement bin boundaries. The algorithm accounts for spheres intersecting a single plane (a face) of the bin, two perpendicular planes (an edge), or three mutually perpendicular planes (a corner). Using this algorithm, comparisons are made against the solid fraction calculated using the more common algorithm in which the solid fraction is determined by assigning all of the sphere's volume to the bin in which the sphere's center of volume is located. A curve fit to the exact algorithm calculations is also given in order to improve computational speed while still maintaining acceptable accuracy. Bin size-to-sphere diameter ratios greater than approximately 30 must be used to give errors less than 5% when using the more traditional method when applied to simple cubic and hexagonal close packing static sphere assemblies. Bin size-to-sphere diameter ratios larger than approximately five are required for random sphere packings. Although time averaged solid fraction measurements are similar using either the exact or center of volume solid fraction schemes, the scatter in the center of volume method is much larger than for the exact method.

See more of this Session: Computational and Numerical Approaches to Particle Flow II
See more of this Group/Topical: Topical 4: Computational Particle Technology

Instability Formation in Granular Taylor-Couette Flows

Brenda Remy, Dept. of Chemical and Biochemical Engineering, Rutgers University, Piscataway, NJ

Thomas M. Canty, Dept. of Chemical and Biochemical Engineering, Rutgers University, Piscataway, NJ

Benjamin Glasser, Dept. of Chemical and Biochemical Engineering, Rutgers University, Piscataway, NJ

Flow instabilities, such as clustering inhomogeneities, can dramatically influence microscopic and macroscopic properties of granular materials. Taylor-Couette flows represent a simple model geometry encompassing both shear and physical boundary interactions - essential ingredients of practical flows. Taylor-Couette fluid flow research in past years has yielded extremely rich information on dynamical behavior. Equivalent steps for granular flows are far from complete.

Instabilities in Taylor-Couette flows of dilute, monodisperse, inelastic granular materials were studied with the use of particle dynamic simulations. The development of density inhomogeneities was observed in pseudo two-dimensional and three-dimensional systems. Fully-developed states were quantified using Fourier methods. For the pseudo two-dimensional case, two main instability modes are observed. The first instability constitutes a density gradient in the radial direction which develops as result of a granular temperature gradient along the inner-outter cylinder gap. As the amount of energy dissipation in the system is increased, a second instability forms in the azimuthal direction analogous to the previously reported particle plug development in one-dimensional flows. When similar flows are studied in a three dimensional system, an additional instability mode develops along the axis. The density inhomogeneities strongly impact macroscopic behavior, as stresses and granular temperatures deviate by up to an order of magnitude from mean values. The influence of solids fraction, material elasticity, surface friction, and gap width on instability formation is also shown.

See more of this Session: Computational and Numerical Approaches to Particle Flow II
See more of this Group/Topical: Topical 4: Computational Particle Technology

Roadmapping of Computational Multiphase Flow

Madhava Syamlal,Office of Research and Development, National Energy Technology Laboratory, U.S. Department of Energy, Morgantown, WV

Chris Guenther, National Energy Technology Laboratory, U.S. Department of Energy, Morgantown, WV

Thomas O'Brien, Simulation, Analysis & Computational, NETL/DOE, Morgantown, WV

In June 2006 several researchers from universities, industry, and national labs met at the National Energy Technology Laboratory (NETL), to discuss outstanding research problems in multiphase flow with particular relevance to energy technologies and to chart out a roadmap for solving such problems. The vision of the workshop was to "ensure that by 2015 multiphase science-based computer simulations play a significant role in the design, operation, and troubleshooting of multiphase flow devices in fossil fuel processing plants." The results of that workshop, including a technology roadmap, were summarized in a workshop report (http://tinyurl.com/c9r7ux). At a follow on workshop held in April 2009 the progress made on the various topics of the roadmap was reviewed, and the process of updating and revising the roadmap was started. The objective of this presentation is to review the roadmap and the progress on various topics relevant to fluid-particle systems, and to involve a wider audience in revising the roadmap and in working on various roadmap topics.

See more of this Session: Industrial Application of Computational and Numerical Approaches to Particle Flow
See more of this Group/Topical: Topical 4: Computational Particle Technology

CFD Simulations of the Fuel Reactor of a Coal-Fired Chemical Looping Combustor AIChE 2009 Fall Proceedings

Kartikeya Mahalatkar[1], ***Thomas O'Brien***[2], *E. David Huckaby*[2] *and John Kuhlman*[3],
[1]*ANSYS-Fluent, Inc., Morgantown, WV,* [2]*Computational Science, NETL/DOE, Morgantown, WV,* [3]*Mechanical Engineering, West Virginia University, Morgantown, WV*

Introduction

Due to its impact on global climate change, the future utilization of coal for power generation will require responsible carbon management (CM) (1). Coal-fired power plants supplied ~49% of the US electricity demand in 2006. Because coal has a higher carbon intensity (C/H ratio ≈ 1.25) than natural gas (C/H ratio = 0.25), which is the other major fossil fuel used for power generation, coal is responsible for ~82% of the CO_2 produced from electrical power generation and 32% of the total CO_2 emissions (2). Although a significant change in the US culture for energy consumption or a rapid growth of alternative power sources could significantly decrease these projections, a 25% increase in coal-fired generation capacity (from 305.2 GW in 2006 to 401.5 GW in 2030) has been projected which implies a corresponding increase in CO_2 emissions. CM involves several steps: 1) separation and capture of a CO_2-rich gas stream; 2) compression and transportation of the CO_2 as a super-critical fluid; and 3) sequestration in a permanent geological reservoir. Of these steps, CO_2 separation is the more costly part of CM, not sequestration as is usually assumed, because of the energy penalty associated with gas separation to obtain a sequestration-ready CO_2 stream.

Chemical Looping Combustion Process

Chemical Looping Combustion (CLC), in which an oxygen carrier is used to provide the oxygen for combustion, inherently creates a separate CO_2 flue gas stream with minimal energy cost. Although this process was initially proposed as a means of achieving greater efficiency (3,4), it was soon recognized (5, 6) that its real strength was in the fact that, as a natural part of the process, it produced a separate stream of combustion products only, CO_2 and H_2O, from which a pure CO_2 stream could be recovered upon compression. However, the process requires two reaction vessels, an Air Reactor (AR) and a Fuel Reactor (FR) and the movement of hot, reactive granular material between them (7). This carrier is oxidized in the AR, separated from the hot, vitiated air stream and transported to the FR where it oxidizes the hydrocarbon fuel, yielding an exhaust gas stream of mainly H_2O and CO_2. The reduced carrier then returns to the air reactor for regeneration, hence the term "looping." These features are represented schematically in Figure 1. In effect, the "oxygen separation" takes place in the cyclone at the exhaust of the AR and is easily effected at low cost due to the natural ease of separating a gas and solid phase. Obviously, the true cost of separation is the capital cost of the two vessel system and the operational cost associated with carrier management.

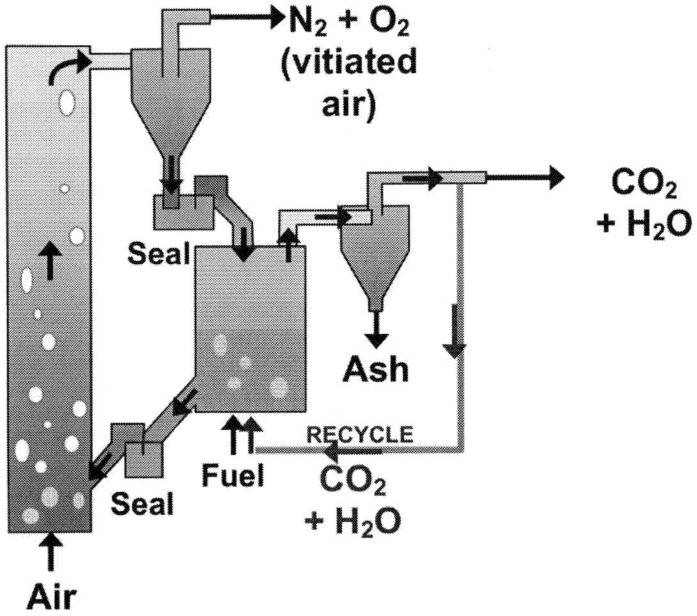

Figure 1. A schematic diagram for the chemical looping combustion of a solid fuel, showing the fuel reactor, cyclone carrier separator, and fuel reactor. A recycled stream of H_2O/CO_2 is required to fluidize the fuel reactor and to facilitate the chemical reactions between the solid carrier and the solid

The net chemical reaction and energy release is identical to that of conventional combustion of the fuel; the Hess's law sum of the reactions in the AR and the FR is just that of the combustion of the fuel directly by O_2. In the AR, the solid coal fuel must be heated by the recycled metal oxide, driving off moisture and volatile material. (This process is probably enhanced in CLC, leading to deeper devolatilization, since the products of devolatilization are removed by reaction with the carrier in the FR, suppressing any effect of back reactions.) The remaining char must be gasified by H_2O (or CO_2), which is recycled from the product stream. The gaseous product of these reactions must then contact the carrier before leaving the bed to obtain complete conversion to H_2O and CO_2. Thus the reactions of the FR involve; 1) the typical reactions involved in coal hydro-gasification; 2) the heterogeneous oxidation of these products by the hot oxygen carrier; and 3) homogeneous gas phase reactions such as the water-gas shift. Further, the reduced carrier particles must be removed from the bed and returned to the air reactor without any accompanying unburned fuel. If unburned fuel is returned to the AR the carbon involved will escape capture and the associated heat release will occur in the AR, however, the over-all operation of the process will not be compromised. Thus, if the operational goal is 90% carbon capture, then 10% return of unburned carbon from the FR to the AR could be tolerated.

A wide variety of oxygen carriers have been proposed and studied (8). The majority of these are based on oxides of transition metals. However, an important exception is the Alstom process based on limestone (9), in which oxygen is carrier by the transformation between

$CaSO_4$ and CaS; CaS is oxidized in the AR to $CaSO_4$ which is then separated and reduced again
by the fuel in the FR. The four requirements usually cited for a carrier are that it must: 1) have
thermodynamic and kinetic carrying capacity; 2) maintain kinetic and mechanical integrity over
many cycles; 3) be cost effect; and 4) be environmentally benign. There have been many studies evaluating these features for many proposed carriers (8,10,11).

CFD Simulations

Until quite recently, work in CLC combustion had been directed toward the utilization of gaseous fuels. However, recently there has been significant work in the CLC of solid fuels (12,13,14). This work is extremely important to the US since we have very large coal reserves.
The data to be simulated were provided by an experimental study of Leion *et al.* (13). The conditions are summarized in Table 1.

Diameter of bed	10 - 30 mm
Height of computational domain	10 m
Flow rate at 0 °C	600 ml/min
Mass of metal oxide particles	20 g
Average diameter of metal particles	150 μm
Average diameter of petcoke particles	210 μm
Average density of metal oxide particles (including) support	4500 kg/m^3
Initial density of petcoke particles	2000 kg/m^3
Initial bed height	35 mm
Initial particle volume fraction	0.6

Table 1 Simulated Fuel Reactor Properties (Leion *et al.*, 2007).

Figure 2 shows the outlet gas concentration at a height of 10 m.

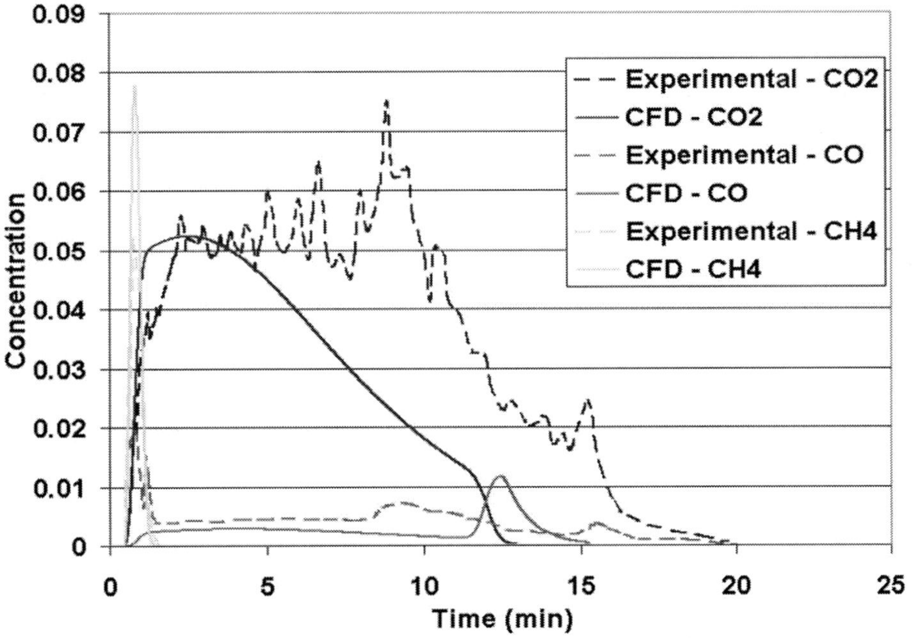

Fig. 6 Concentration of CO_2, CO and CH_4 at a height of 10 m above the bed.

The peak CH_4 concentration predicted by the simulations differs by 30% in comparison with experimental values. This difference may be attributed to uncertainties of the experimental conditions related to coal feeding (15).

References

1. Hansen, J., M. Sato, J. Glascoe, and R. Ruedy (1998), "A common-sense climate index: Is climate changing noticeably?" *Proc. Natl. Acad. Sci.* **95**, 4113-4120.
2. Annual Energy Outlook 2007, DOE/EIA-0383 (2007).
3. Knoche, K.F., and H. Richter (1968), "Verbesserung der Reversibilität von Verbrennungsprozessen," Brennstoff-Warme-Kraft, **20**, 205-210.
4. Richter, H.J. and K. Knoche (1983), "Reversibility of combustion processes," Efficiency and Costing, Second Law Analysis of Processes, ACS Symp. **235**, Washingto D.C., pp.71-85.
5. Ishida, M., and H. Jin (1994), "A new advanced power-generation system using chemical-looping combustion," *Energy*, **19**, 415-422.
6. Lyngfelt, A. and B. Leckner (1999), "Technologies for CO_2 separation," *Minisymposium on Carbon Dioxide Capture and Storage*, Chalmers University of Technology and Göteborg University, Göteborg, October 22, 1999.
7. Lyngfelt, A., B. Leckner, and T. Mattisson (2001)," A fluidized-bed combustion process with inherent CO_2 separation; application of chemical-looping combustion", *Chem. Eng. Sci.*, **56**, 3101-3113.
8. Johansson, M., T. Mattisson, M. Rydén, and A. Lyngfelt (2006), "Carbon Capture via

Chemical-Looping Combustion and Reforming," *International Seminar on Carbon Sequestration and Climate Change*, Rio de Janeiro, Brazil, October 24 – 27, 2006.

9. Andrus, H.E., G. Burns, J.H. Chiu, G.N. Liljedahl, P.T. Stromberg, and P.R. Thibeault (2008), "Hybrid Combustion-Gasification Chemical Looping" Alstom Final Report Phase III, DE-FC26-03NT41866, US DOE NETL, Sept. 30, 2008.
10. Adánez, J.; L.F. de Diego, F. García-Labiano, F.; P. Gayán, A. Abad (2004), *Energy Fuels*, **18**, 371-377.
11. Adánez, J.; F. García-Labiano, L.F. de Diego, P. Gayán, A. Abad, and J. Celaya (2005), Development of oxygen carriers for chemical-looping combustion. In *Carbon Dioxide Capture for Storage in Deep Geologic Formations - Results from the CO_2 Capture Project*. Thomas, D., Benson, S., Eds.; Elsevier Ltd.: Oxford, UK, 2005; Vol. 1, Chapter 34.
12. Mattisson, T., J.C. Abanades , A. Lyngfelt, A. Abad, M. Johansson, J. Adánez, F. García-Labiano, L.F. de Diego, P. Gayán, B. Kronberger, H. Hofbauer, M. Luisser, J.M. Palacios, D. Alvares, G. Grasa, J. Oakey, B. Arias, M. Orjala, and V.-P. Heiskanen, "Capture of CO_2 in Coal Combustion (CCCC)", Final report, ECSC Coal RTD Programme, Contract No: 7220-PR-125, Chalmers University of Technology, 2005.
13. Leion, H., T. Mattisson and A. Lyngfelt (2007), "The use of petroleum coke as fuel in chemical-looping combustion," *Fuel*, **86**, 1947–1958.
14. Mattisson, T., A. Lyngfelt, and H. Leion (2009), "Chemical-looping with oxygen uncoupling for combustion of solid fuels," *Inter. J. Greenhouse Gas Control*, **3**, 11-19.
15. Personal communication, H. Leion, Sept. 24, 2009.

Cratering of a Particle Bed by a Turbulent Subsonic Jet

Casey Q. LaMarche,Chemical Engineering, University of Florida, Gainesville, FL

Jennifer S. Curtis, Chemical Engineering, University of Florida, Gainesville, FL

Aldo G. Benavides M., Applied Mechanics, Chalmers University of Technology, 412 96 Göteborg, Sweden

Berend van Wachem, Mechanical Engineering, Imperial College London, London SW7 2AZ, England

Future space exploration necessitates improved gas-solid and solid-solid interaction models. Impinging rocket plumes from spacecraft interact with the lunar regolith surface and release a high velocity particle-spray that is potentially hazardous to surrounding surface structures and surface architecture. Experiments performed at NASA-KSC provide data on the cratering of a particle bed by a turbulent subsonic jet of gas. In order to model this system, an Eulerian treatment is employed using locally averaged equations of motion along with closure relations describing the solids-phase stress derived using concepts based on dense-phase gas kinetic theory. The solids phase frictional stresses, which occur at high solids volume fractions and result from sustained particle-particle contacts, are based on critical state theory. Standard commercial codes employ these frictional stress models, but poorly predict the cratering dynamics found experimentally and predict a particle bed that is too liquid-like. However, a continuum frictional stress model in which the normal frictional pressure and frictional viscosity is increased significantly improves the predictions for the cratering behavior. In addition, the effect of the changes in the jet velocity and density are explored and compared with experimental data for crater width and depth.

See more of this Session: Industrial Application of Computational and Numerical Approaches to Particle Flow
See more of this Group/Topical: Topical 4: Computational Particle Technology

Solution for the Speed of Decompression Waves in Fluidization

B. A. Kashiwa,Theoretical Division, LANL/NNSA/DOE, Los Alamos, NM

We consider the decompression wave in fluidization, as described by Wallis, et al. [Int. J. Multiphase Flow,19, 839--874, 1993], and give the analytic solution for the one--dimensional wave speed. In its simplest form, the decompression wave is a concentration wave moving upward into a close--packed region. The wave structure is such that this wave connects the close--packed region with a nonpacked, expanding, region that is in local equilibrium the the hydrostatic pressure gradient. The solution is general; hence it is valid for beds of solid grains fluidized by either gas or liquid, sedimentation, and for small gas bubbles rising in a liquid. Averaged two--body forces are the main physical effects that determine the wave speed. The hyperbolic model equations developed by Kashiwa & Rauenzahn [Proc. 3rd Int. Symp. on Two--Phase Flow Modeling and Exp., Edizioni ETS, 2004] supply the general framework; data from experiments by DiFelice [in Wallis, et al. 1993] provide an undetermined coefficient associated with the viscous part two--body force. The solution is shown to be capable of predicting the onset of bubbling in the fluidization of solid grains, by comparison with the data of Rietema [Chem. Eng. Sci., 37, 1125--1150, 1982].

See more of this Session: Industrial Application of Computational and Numerical Approaches to Particle Flow
See more of this Group/Topical: Topical 4: Computational Particle Technology

Using a CFD-DEM Modeling Approach to Address Upstream Oil Production Challenges

André Leibson Martins, Cenpes/Pdp/Tep, Petrobras S.A., Rio De Janeiro, Brazil

João Vicente M. Magalhães, Cenpes/Pdp/Tep, Petrobras S.A., Rio De Janeiro, Brazil

Alex Tadeu A. Waldmann, Cenpes/Pdp/Tep, Petrobras S.A., Rio De Janeiro, Brazil

Rosana Fátima T. Lomba, Cenpes/Pdp/Tep, Petrobras S.A., Rio De Janeiro, Brazil

Jairo Zago de Souza, Engineering Simulation and Scientific Software LTDA., Rio De Janeiro, Brazil

João Américo Aguirre Oliveira Jr., Engineering Simulation and Scientific Software LTDA., Rio De Janeiro, Brazil

Daniel da Cunha Ribeiro, Engineering Simulation and Scientific Software LTDA., Rio De Janeiro, Brazil

Wai Sam Wong, DEM Solutions Ltd., Edinburgh, United Kingdom

Richard D. LaRoche,DEM Solutions USA Inc., Lebanon, NH

During horizontal oil well completion, there is an important stage that guarantees the integrity and operation during the oil production known as gravel packing. This operation allows an annulus to be formed between the well walls and production column which can be filled by particulate material, frequently sand. When the operation is not controlled, there is a problem called "premature filling" which can impair well production. Another important issue that should be adequately addressed is the existence of natural fractures found during drilling and completion of hydrocarbons wells. These fractures may drain the drilling fluid and prevent controlled production. Looking for better understanding of these fluid-particle flow processes, Petrobras, ESSS and DEM Solutions are developing a new simulation approach based on the combination of computational fluid dynamics (CFD) and discrete element model (DEM) techniques. This presentation provides details on the modeling methodology and simulation results using ANSYS-FLUENT® and EDEM® software packages.

See more of this Session: Industrial Application of Computational and Numerical Approaches to Particle Flow
See more of this Group/Topical: Topical 4: Computational Particle Technology

AUTHOR INDEX